Undergraduate Texts in Mathematics

Editors
J.H. Ewing
F.W. Gehring
P.R. Halmos

Undergraduate Texts in Mathematics

(continued after Index)

Gerald A. Edgar

Measure, Topology, and Fractal Geometry

With 96 Illustrations and 16 Color Plates

514.74
E

Springer-Verlag
New York Berlin Heidelberg
London Paris Tokyo Hong Kong

Gerald A. Edgar
Department of Mathematics
The Ohio State University
Columbus, OH 43210-1174
USA

Cover illustration is a computer-generated fractal (plate 11).

Illustration (p. 51) from *The Mouse and His Child* by Russel Hoban, pictures by Lillian Hoban. Illustration Copyright © 1967 by Lillian Hoban. Reprinted by permission of Harper & Row, Publishers, Inc., and of Faber and Faber Ltd.

Two illustrations (p. 148) from *Geometric Measure Theory* by Frank Morgan, Copyright © 1988 by Academic Press, Inc. Reprinted by permission of Academic Press and Frank Morgan.

Library of Congress Cataloging-in-Publication Data
Edgar, Gerald A., 1949–
    Measure, topology, and fractal geometry/Gerald A. Edgar.
      p.   cm.—(Undergraduate texts in mathematics)
    Includes bibliographical references.
    ISBN 0-387-97272-2 (alk. paper)
    1. Fractals. 2. Measure theory. 3. Topology. I. Title.
  II. Series.
  QA614.86.E34   1990
  514′.74—dc20   90-33060

Printed on acid-free paper.

Typeset with $A_{M}S$-TEX, the TEX macro system of the American Mathematical Society.
Printed and bound by R. R. Donnelley & Sons, Harrisonburg, Virginia.
Printed in the United States of America.

9 8 7 6 5 4 3 2 1

ISBN 0-387-97272-2 Springer-Verlag New York Berlin Heidelberg
ISBN 3-540-97272-2 Springer-Verlag Berlin Heidelberg New York

Preface

What is a fractal? Benoit Mandelbrot coined the term in 1975. There is (or should be) a mathematical definition, specifying a basic idea in geometry. There is also a figurative use of the term to describe phenomena that approximate this mathematical ideal. Roughly speaking, a fractal set is a set that is more "irregular" than the sets considered in classical geometry. No matter how much the set is magnified, smaller and smaller irregularities become visible. Mandelbrot argues that such geometric abstractions often fit the physical world better than regular arrangements or smooth curves and surfaces. On page 1 of his book, *The Fractal Geometry of Nature*, he writes, "Clouds are not spheres, mountains are not cones, coastlines are not circles, and bark is not smooth, nor does lightning travel in a straight line."[36, p. 1]*

To define **fractal**, Mandelbrot writes: "A fractal is by definition a set for which the Hausdorff-Besicovitch dimension strictly exceeds the topological dimension."[36, p. 15] It might be said that this book is a meditation on that verse. Study of the Hausdorff dimension requires measure theory (Chapter 5); study of topological dimension requires metric topology (Chapter 2). Note, however, that Mandelbrot later expressed some reservations about this definition: "In science its [the definition's] generality was to prove excessive; not only awkward, but genuinely inappropriate ... This definition left out many 'borderline fractals', yet it took care, more or less, of the frontier 'against' Euclid. But the frontier 'against' true geometric chaos was left wide open."[46, p. 159] We will discuss in Chapter 6 a way (proposed by James Taylor) to repair the definition. Mandelbrot himself, in the second printing of [36, p. 459], proposes "... to use 'fractal' without a pedantic definition, to use 'fractal dimension' as a generic term applicable to *all* the variants in Chapter 39, and to use in each specific case whichever definition is the most appropriate." I have not adopted the first of these suggestions in this book, since a term without a "pedantic definition" cannot be discussed mathematically. I have, however, used the term "fractal dimension" as suggested.

---

*Bracketed numbers like this refer to the references collected on page 223.

This is a *mathematics* book. It is not about how fractals come up in nature; that is the topic of Mandelbrot's book [**36**]. It is not about how to draw fractals on your computer. (I did have a lot of fun using a Macintosh to draw the pictures for the book, however. There will be occasional use of the Logo programming language for illustrative purposes.) Complete proofs of the main results will be presented, whenever that can reasonably be done. For some of the more difficult results, only the easiest non-trivial case of the proof (such as the case of two dimensions) is included here, with a reference to the complete proof in a more advanced text.

The main examples that will be considered are subsets of Euclidean space (in fact, usually two-dimensional Euclidean space); but as we will see, it is helpful to deal with the more abstract setting of "metric spaces".

This book deals with *fractal geometry*. It does not cover, for example, chaotic dynamical systems. That is a separate topic, although it is related. Another book of the same size could be written* on it. This book does not deal with the Mandelbrot set. Some writing on the subject has left the impression that "fractal" is synonymous with "Mandelbrot set"; that is far from the truth. This book does not deal with stochastic (or random) fractals; their rigorous study would require more background in probability theory than I have assumed for this book.

Prerequisites. (1) Experience in reading (and, preferably, writing) mathematical proofs is essential, since proofs are included here. I will say "necessary and sufficient" or "contrapositive" or "proof by induction" without explanation. Readers without such experience will only be able to read the book at a more superficial level by skipping many of the proofs. (Of course, no mathematics student will want to do that!)

(2) Basic abstract set theory will be needed. For example, the abstract notion of a function; finite vs. infinite sets; countable vs. uncountable sets.

(3) The main prerequisite is calculus. For example: What is a continuous function, and why do we care? The sum of an infinite series. The limit of a sequence. The least upper bound axiom (or property) for the real number system. Proofs of the important facts are included in many of the modern calculus texts, but unfortunately they are often omitted by the instructor (because of student resistance or simple lack of time).

Advice for the reader. Here is some advice for those trying to read the book without guidance from an experienced instructor. The most difficult (and tedious) parts of the book are probably Chapters 2 and 5. But these chapters lead directly to the most important parts of the book, Chapters 3 and 6. Most of Chapter 2 is independent of Chapter 1, so to ease the reading of Chapter 2, it might be reasonable to intersperse parts of Chapter 2 with parts of Chapter 1.

---

*For example, [**14**].

In a similar way, Chapters 3, 4, and 5 are mostly independent of each other, so parts of these three chapters could be interspersed with each other.

There are many exercises scattered throughout the text. Some of them deal with examples and supplementary material, but many of them deal with the main subject matter at hand. Even though the reader already knows it, I must repeat: Understanding will be greatly enhanced by work on the exercises (even when a solution is not found). Answers to some of the exercises are given elsewhere in the book, but in order to encourage the reader to devote more work to the exercises, I have not attempted to make them easy to find. When an exercise is simply a declarative statement, it is to be understood that it is to be proved. (Professor Ross points out that if it turns out to be false, then you should try to salvage it.) Some exercises are easy and some are hard. I have even included some that I do not know how to solve. (No, I won't tell you which ones they are.)

Take a look at the Appendix. It is intended to help the reader of the book. There is an index of the main terms defined in the book; an index of notation; and a list of the fractal examples discussed in the text.

Some illustrations are not referred to in the main text. An instructor who knows what they are may use them for class assignments at the appropriate times.

Some of the sections that are more difficult, or deal with less central ideas, are marked with an asterisk (*). They should be considered optional. A section of "Remarks" is at the end of each chapter. It contains many miscellaneous items, such as: references for the material in the chapter; more sophisticated proofs that were omitted from the main text; suggestions for course instructors.

Notation. Most notation used here is either explained in the text, or else taken from calculus and elementary set theory. A few reminders and additional explanations are collected here.

Integers: $\mathbf{Z} = \{\cdots, -2, -1, 0, 1, 2, \cdots\}$.

Natural numbers or positive integers: $\mathbf{N} = \{1, 2, 3, \cdots\}$.

Real numbers: $\mathbf{R} = (-\infty, \infty)$.

Intervals of real numbers:

$$(a, b) = \{x : a < x < b\};$$
$$(a, b] = \{x : a < x \leq b\}; \quad \text{etc.}$$

The notation (a, b) also represents an ordered pair, so the context must be used to distinguish.

Set difference or relative complement: $X \setminus A = \{x \in X : x \notin A\}$.

If $f : X \to Y$ is a function, and $x \in X$, I will use parentheses $f(x)$ for the value of the function at the point x; if $C \subseteq X$ is a set, I will use square brackets for the image set $f[C] = \{f(x) : x \in C\}$.

The union of a family $(A_i)_{i \in I}$ of sets, written

$$\bigcup_{i \in I} A_i,$$

consists of all points that belong to at least one of the sets A_i. The intersection

$$\bigcap_{i \in I} A_i$$

consists of the points that belong to all of the sets A_i. The family $(A_i)_{i \in I}$ is said to be **disjoint** iff $A_i \cap A_j = \emptyset$ for any $i \neq j$ in the index set I.

The **supremum** (or least upper bound) of a set $A \subseteq \mathbf{R}$ is written $\sup A$. By definition $u = \sup A$ satisfies (1) $u \geq a$ for all $a \in A$, and (2) if $y \geq a$ for all $a \in A$, then $y \geq u$. Thus, if A is not bounded above, we write $\sup A = \infty$, and if $A = \emptyset$, we write $\sup A = -\infty$. The **infimum** (or greatest lower bound) is $\inf A$. The **upper limit** of a sequence $(x_n)_{n=1}^{\infty}$ is

$$\limsup_{n \to \infty} x_n = \lim_{n \to \infty} \sup_{k \geq n} x_k.$$

And, if $\alpha(r)$ is defined for real $r > 0$,

$$\limsup_{r \to 0} \alpha(r) = \lim_{s \to 0} \sup_{0 < r < s} \alpha(r).$$

Similar notation is used for the **lower limit** or $\liminf$.

The sign ☺ signals the end of a proof.

The origin of the book. I offered a course in 1987 at The Ohio State University on fractal geometry. It was intended for graduate students in mathematics, and it was based on the books of Hurewicz and Wallman [30] and Falconer [19]. I tried to keep the prerequisites at a low enough level that, for example, a graduate student in physics could take the course. The prerequisites listed were: metric topology and Lebesgue measure.* When the course was announced, I began getting inquiries from many other types of students, who were interested in studying fractal geometry more rigorously, but did not have even this minimal background. For example, a student in computer science with a strong background in calculus would still have required two more years of mathematics study ("Advanced Calculus" and "Introductory Real Analysis") before being prepared for the course. This book is intended to fit this sort of student. Only a small part of those two courses is actually required for the study of fractal geometry, at least at the most elementary level. The required topics from metric

---

*Then I found, to my surprise, that Lebesgue integration is not considered necessary for physics students. I suppose the fact that I find this incredible is an illustration of my ignorance of how mathematics is applied in practice.

topology and measure theory are covered in Chapters 2 and 5. (Mathematics students may be able to skip much of these two chapters.)

This book is directly derived from notes prepared for use in a course offered in 1988 in connection with the program for talented high school students that is run here at The Ohio State University by Professor Arnold Ross for eight weeks every summer. The influence of these young students can be seen in many small ways in the book. (In particular, 1.5.7 and 1.6.1.) Past practice in the Ross program suggested the fruitful use of ultrametric spaces.

Parts of the manuscript were read by Manav Das, Don Leggett, William Mc-Worter, Lorraine Rellick, and Karl Schmidt. Their comments led to many improvements in the manuscript. I would like to thank the many people at Springer-Verlag New York, especially mathematics editor Rüdiger Gebauer, mathematics assistant editor Susan Gordon, mathematics editor Ulrike Schmickler-Hirzebruch (editor of the "Undergraduate Texts in Mathematics" series), and production editor Susan Giniger.

The book was prepared by the author on an Apple Macintosh Plus provided by the Department of Mathematics at The Ohio State University. The text was processed using the TEX system (written by Donald Knuth), and the $\mathcal{AMS}$-TEX macro package. The AMS fonts were very useful, in addition to the Computer Modern fonts. TEX was in the form of the public domain programs ctex by Tomas Rokicki and dvi2ps by Mark Senn and others, both adapted to run under the MPW Shell. The pictures were prepared on the Macintosh using Coral Object Logo, LCSI Logo, MacDraw II, and the Scrapbook. Camera-ready copy was produced on an Apple LaserWriter; the color plates are from color separations produced by MacDraw II.

G. A. Edgar

Columbus, Ohio
March 6, 1990

His days and times are past,
And my reliances on his fracted dates
Have smit my credit: I love and honor him
—W. Shakespeare, *Timon of Athens*

Contents

1

Fractal Examples

A few basic mathematical examples of fractals will be introduced in this chapter. Their analysis, and especially the question of what makes them "fractals", must be postponed until much later in the book.

One of the surprising ideas in the subject is the contention that the "dimension" of a set might be a real number that is not an integer. If we say that a set C in the plane has dimension 1.7, then we mean that its properties are "between" those of a curve (dimension 1) and an open region (dimension 2). The technical aspects of such "fractal" dimensions are discussed in Chapter 6.

Another characteristic feature of fractals emphasizes their difference from the sets of classical geometry. Typically, a fractal looks irregular; but more importantly, after it is magnified it still looks irregular. A typical set from classical geometry becomes very simple looking if it is magnified enough. One of the ways in which this behavior under magnification can be specified is through the use of an iterated function system. Several of the examples in this chapter have descriptions in terms of iterated function systems, but the more detailed discussion is in Chapter 4.

1.1. THE TRIADIC CANTOR DUST

We will begin with an example. It will be treated in several different ways. Examples that we study later will often have something in common with one or more of the aspects treated for this example. This example is known as "the Cantor set". Mandelbrot has called it the "Cantor dust"; this descriptive name shows what kind of set it is. (The term "dust" refers to the fact that the set is zero-dimensional; this will be discussed in Chapter 3.)

Construction by tremas. The **triadic Cantor dust** is a subset of the line $\mathbb{R}$. A sequence of approximations is first defined. Start with the closed interval $C_0 = [0, 1]$. Then the set C_1 is obtained by removing the "middle third" from

$[0, 1]$, leaving $[0, 1/3] \cup [2/3, 1]$. The next set C_2 is defined by removing the middle third of each of the two intervals of C_1. This leaves

$$C_2 = [0, 1/9] \cup [2/9, 1/3] \cup [2/3, 7/9] \cup [8/9, 1].$$

And so on. (See Figure 1.1.1.) The **triadic Cantor dust** is the "limit" C of the sequence C_n of sets. The sets decrease: $C_0 \supseteq C_1 \supseteq C_2 \supseteq \cdots$. So we will define the "limit" to be the intersection of the sets,

$$C = \bigcap_{k \in \mathbb{N}} C_k.$$

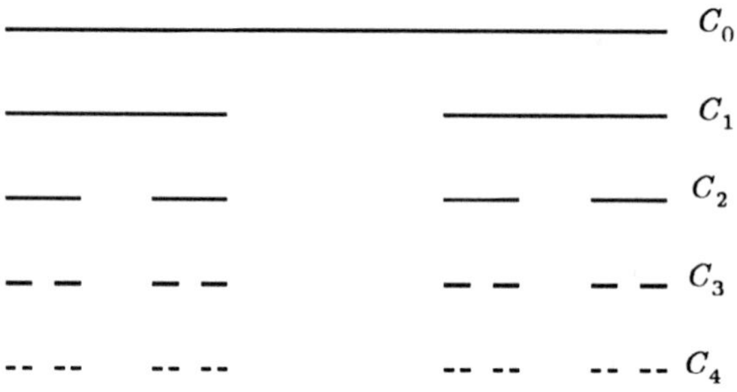

Figure 1.1.1. The triadic Cantor dust.

We will later do other constructions in a similar way. The parts that are removed are called **tremas**.

The sequence of sets is defined recursively. This means that it will often be easy to prove facts about the sets by induction. For example, the set C_k consists of 2^k disjoint closed intervals, each of length $(1/3)^k$. So the **total length of C_k**, the sum of the lengths, is $(2/3)^k$. The limit is

$$\lim_{k \to \infty} \left(\frac{2}{3} \right)^k = 0.$$

So the total length of the Cantor dust itself is zero. (The mathematical version of "total length" is called "Lebesgue measure". It will be dealt with in Chapter 5). So total length is not a very useful way to compute the size of C. We will see that this is related to the fact that the fractal dimension of C is < 1.

Let us consider more carefully what points constitute the Cantor set. If $[a, b]$ is one of the closed intervals that makes up one of the approximations C_k, then the endpoints a and b belong to *all* of the future sets C_m, $m \geq k$, and therefore belong to the intersection C. (Again, prove this by induction.) Taking all the endpoints of all the intervals of all the approximations C_k, we get an infinite set of points, all belonging to C. (This set of endpoints is only a countable set, however.)

But it is important to note that there are points in C other than these endpoints.

(1.1.2) EXERCISE. *The point 1/4 is not an endpoint of any interval of any set C_k. But the point 1/4 belongs to C.*

It is also important to note that C is not like the usual sets of elementary geometry. At first, it is likely to tax your powers of geometrical visualization. If you can't imagine it today, try again tomorrow. Here are a few tidbits to help:

(1.1.3) EXERCISE.

(1) *The set C contains no interval (of positive length).*
(2) *The set C has no isolated points: that is, if $a \in C$, then for every $\varepsilon > 0$, no matter how small, the interval $(a - \varepsilon, a + \varepsilon)$ contains points of C in addition to a.*
(3) *The set C is closed: that is, if $a \in \mathbb{R}$ has the property that every interval of the form $(a - \varepsilon, a + \varepsilon)$ intersects C, then $a \in C$.*

(1.1.4) EXERCISE. *Let r be a positive number. How many tremas are there with length $\geq r$?*

Coordinates. There is a convenient way to characterize the elements of the triadic Cantor dust in terms of their expansions in base 3.

First we will review the standard facts concerning base 3. You know how expansions in the usual base 10 work. Base 3 is of course similar. Every positive integer x has a unique representation

$$x = \sum_{j=0}^{M} a_j 3^j,$$

where the "digits" a_j are chosen from the list $0, 1, 2$. For example,

$$15 = 0 \cdot 3^0 + 2 \cdot 3^1 + 1 \cdot 3^2.$$

We will sometimes write simply $15 = (120)_3$. It is understood that the subscript specifying the base is always written in base ten.

Similarly, we may represent fractions: Every number x between 0 and 1 has a representation in the form

$$x = \sum_{j=-\infty}^{-1} a_j 3^j,$$

with digits $0, 1, 2$. These are written with a "radix point" (or "ternary point" in this case):

$$7/9 = (0.21)_3,$$
$$1/4 = (0.02020202 \cdots)_3 \qquad \text{(a repeating expansion)},$$
$$\sqrt{2} = (1.1020112 \cdots)_3 \qquad \text{(non-repeating)}.$$

Some numbers (rational numbers of the form $a/3^k$) admit two different expansions. for example,

$$1/3 = (0.1000000\cdots)_3 = (0.0222222\cdots)_3.$$

(1.1.5) PROPOSITION. *Let $x \in [0, 1]$. Then x belongs to the triadic Cantor dust C if and only if x has a base 3 expansion using only the digits 0 and 2.*†

PROOF: The first place to the right of the ternary point is a 1 if and only if x is between

$$(0.1000000\cdots)_3 = 1/3 \quad \text{and} \quad (0.1222222\cdots)_3 = 2/3.$$

The first trema is the interval $(1/3, 2/3)$. After this trema is removed, we have C_1. (The numbers $1/3$ and $2/3$ each have two expansions, one with a 1 in the first place, and one without. So they should not be removed.) So C_1 contains exactly the numbers in $[0, 1]$ that have a base 3 expansion not using 1 in the first place. The second place of a number x in C_1 is a 1 if and only if x belongs to one of the second-level tremas $(1/9, 2/9)$ or $(7/9, 8/9)$. When these tremas are removed, we have C_2. So C_2 contains exactly the numbers in $[0, 1]$ that have a base 3 expansion not using 1 in the first or second place. Continuing in this way, we see that the points remaining in $C = \bigcap_{k \in \mathbb{N}} C_k$ are exactly the numbers in $[0, 1]$ that have a base 3 expansion not using 1 at all. ☺

The Cantor dust is uncountable. This follows from the representation just proved, together with the observation that each real number has at most two representations base 3. (Actually, for numbers in the Cantor dust, two different sequences of 0's and 2's always represent different real numbers. See Exercise 1.6.7.)

Construction by translations. Suppose L is a subset of $\mathbb{R}$ and s is a real number. The **translate** of L by s is the set

$$\{x + s : x \in L\}.$$

That is, we add s to each element of L. This is sometimes written $L + s$.

We construct recursively a sequence (L_k) of subsets of the line $\mathbb{R}$, together with a sequence (s_k) of real numbers. (Figure 1.1.6.) Begin with the number $s_0 = 2/3$, and a starting set L_0 consisting of the one point 0. The next set L_1 is obtained by combining L_0 with its translate by s_0. So $L_1 = \{0, 2/3\}$. The next term in the sequence (s_k) will be $1/3$ times the previous one; $s_1 = (1/3)(2/3) = 2/9$. It will be used to obtain L_2 from L_1:

$$L_2 = L_1 \cup (L_1 + s_1) = \{0, 2/9, 2/3, 8/9\}.$$

---

†This proposition will show that $1/4 \in C$, which is part of Exercise 1.1.2. But I hope you already solved it yourself. Was your solution for the special number $1/4$ simpler than this general case?

L_0 .

L_1 .   .

L_2 .   .   . .

L_3 . . . .

L_4 .. ..

Figure 1.1.6. Translation construction.

Then $s_2 = (1/3)s_1$ and $L_3 = L_2 \cup (L_2 + s_2)$. And so on. The set of interest is the "limit" L of the sequence L_n. This sequence is increasing: $L_0 \subseteq L_1 \subseteq L_2 \subseteq \cdots$, so one reasonable definition for a limit would be the union:

$$L = \bigcup_{k \in \mathbb{N}} L_k.$$

But in fact, we will see later another (and, for our purposes, better) way to define the "limit" of a sequence of sets such as (L_k).

What is the connection between this construction and the previous construction of the Cantor dust C? The set L_k consists of 2^k points. They are exactly the left endpoints of the intervals that make up the set C_k. Or, they are the right endpoints of the tremas removed from $[0, 1]$ to construct C_k (plus the one point 0).

The points of L_k are the numbers in $[0, 1]$ that have a base 3 representation with k digits involving only 0's and 2's. (Prove this by induction on k.) For example, L_2 consists of

$$(0.00)_3 = 0,$$
$$(0.02)_3 = 2/9,$$
$$(0.20)_3 = 2/3,$$
$$(0.22)_3 = 8/9.$$

Certainly the set $L = \bigcup_{k \in \mathbb{N}} L_k$ is not equal to the Cantor dust C. The number 1/4 does not belong to L. But L is "close" to C in the following sense.

(1.1.7) PROPOSITION. *If $x \in C$, then x is the limit of a sequence of points of L.*

PROOF: Since $x \in C$, we know that x has a base 3 representation:

$$x = \sum_{j=1}^{\infty} a_j 3^{-j}, \qquad \text{each } a_j = 0 \text{ or } 2.$$

If this representation is truncated after only k terms, we get a number

$$x_k = \sum_{j=1}^{k} a_j 3^{-j},$$

which is an element of L_k. Now

$$|x - x_k| = \sum_{j=k+1}^{\infty} a_j 3^{-j} \leq \sum_{j=k+1}^{\infty} 2 \cdot 3^{-j} = 3^{-k}.$$

Since $\lim_{k \to \infty} 3^{-k} = 0$, we may conclude that $\lim_{k \to \infty} x_k = x$. Thus $x \in C$ is the limit of the sequence $x_k \in L$. ☺

The set L is **dense** in the Cantor set C. This means that $L \subseteq C$, and every point of C is the limit of a sequence of points of L.

Iterated function system. Let $r > 0$ and $a \in \mathbb{R}$. The **dilation** on $\mathbb{R}$ with **ratio** r and **center** a is the function $f: \mathbb{R} \to \mathbb{R}$ given by $f(x) = rx + (1 - r)a$.
Consider the two dilations on $\mathbb{R}$ defined by:

$$f_1(x) = \frac{x}{3}, \qquad f_2(x) = \frac{x + 2}{3}.$$

They both have ratio $1/3$. The first has center 0 and the second has center 1.

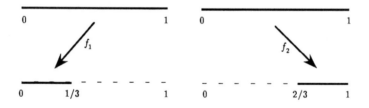

Figure 1.1.8. Two dilations.

(1.1.9) PROPOSITION. *The triadic Cantor dust C satisfies*

$$C = f_1[C] \cup f_2[C].$$

PROOF: It follows by induction that $C_{k+1} = f_1[C_k] \cup f_2[C_k]$ for $k = 0, 1, \cdots$.
First I will prove that $C \subseteq f_1[C] \cup f_2[C]$. Suppose $x \in C$. Then $x \in C_1$. So either $x \in [0, 1/3]$ or $x \in [2/3, 1]$. We take the case in which $x \in [2/3, 1]$; the

other case is similar. Now for any k, we know $x \in C_{k+1} = f_1[C_k] \cup f_2[C_k]$. But $f_1[C_k] \subseteq f_1[[0,1]] = [0,1/3]$, so in fact $x \in f_2[C_k]$, or $3x - 2 \in C_k$. This is true for all k, so $3x - 2 \in \bigcap_{k\in\mathbf{N}} C_k = C$. Thus, $x \in f_2[C]$. In the other case, $x \in f_1[C]$. So in any case, we have $x \in f_1[C] \cup f_2[C]$.

Next I will prove that $C \supseteq f_1[C] \cup f_2[C]$. Suppose $x \in f_1[C] \cup f_2[C]$. Either $x \in f_1[C]$ or $x \in f_2[C]$. We take the case $x \in f_2[C]$; the other case is similar. Thus, $3x - 2 \in C$. Now for any k, we know $3x - 2 \in C_k$, or $x \in f_2[C_k] \subseteq C_{k+1}$. Thus $x \in \bigcap_{k\in\mathbf{N}} C_{k+1} = \bigcap_{k\in\mathbf{N}} C_k = C$. This completes the proof that $C \supseteq f_1[C] \cup f_2[C]$. ☺

We will call the pair (f_1, f_2) an **iterated function system**, and we will say that C is the* **invariant set** (or **attractor**) of the iterated function system (f_1, f_2).

(1.1.10) EXERCISE. *There are sets $A \neq C$ also satisfying $A = f_1[A] \cup f_2[A]$. How many can you find?*

1.2. THE SIERPIŃSKI GASKET

The next example is a set in the plane known as the **Sierpiński gasket**.

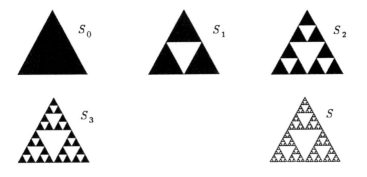

Figure 1.2.1. The Sierpiński gasket.

Construction by tremas. Start with an equilateral triangle (including its inside) with side length 1. Call it S_0. It may be subdivided into four smaller triangles, using lines joining the midpoints of the sides. The smaller triangles have side length $1/2$. The middle triangle is rotated 180 degrees compared to the others. The trema to be removed is the middle triangle (but not its boundary). After it is removed, the remaining set is S_1, a subset of S_0. Now each of the three remaining triangles should be subdivided into smaller triangles with edge length $1/4$, and the three middle triangles removed. The result is S_2, a subset of S_1. We should continue in the same way, to obtain a sequence S_k of sets. The

---

*In Chapter 4, extra conditions will be added that will make the invariant set of an iterated function system unique, justifying the word "the".

Sierpiński gasket is the limit S of this sequence of sets. (Figure 1.2.1.) The sequence is decreasing ($S_0 \supseteq S_1 \supseteq S_2 \supseteq \cdots$), so by the "limit" we mean the intersection $S = \bigcap_{k \in \mathbb{N}} S_k$.

The set S_k consists of 3^k triangles, with side 2^{-k}. So the **total area** of S_k is $3^k \cdot (2^{-k})^2 \cdot \sqrt{3}/4$. This converges to 0 as $k \to \infty$. The total area of the Sierpiński gasket itself is therefore 0. Thus "area" is not very useful in measuring the size of S. Area is used to measure the size of a set of dimension 2. A line segment, which has dimension 1, has area 0. In a similar way, we will see that the Sierpiński gasket S can be said to have dimension less than 2.

The line segments that make up the boundary of one of the triangles of S_n remain in all the later approximations S_k, $k \geq n$. So the set S contains at least all of these line segments. In S_k there are 3^k triangles, each having 3 sides of length 2^{-k}. So the "total length" of S is at least $3^k \cdot 3 \cdot 2^{-k}$. This goes to ∞ as $k \to \infty$. So it makes sense to say that the total length of S is infinite. So "length" is not very useful to measure the size of S. Length is used to measure the size of a set of dimension 1. A square (with its inside), which has dimension 2, has infinite length, since it contains as many disjoint line segments as you like. In a similar way, we will see that the Sierpiński gasket S can be said to have dimension greater than 1.

So S supposedly has dimension greater than 1 but also less than 2. There is no integer between 1 and 2. The way around this dilemma, proposed by Hausdorff in 1919, is to allow the dimension of a set to be a fraction. According to Hausdorff's definition (Chapter 6), the Sierpiński gasket has dimension approximately 1.58.

Iterated function system. Let $r > 0$ be a real number and let a be a point in the plane. The **dilation** with ratio r and center a is a map f of the plane to itself such that each point x is mapped to a point $f(x)$, which is on the ray from a through x, and the distance from a to $f(x)$ is r times the distance from a to x. By convention, $f(a) = a$ also. (Figure 1.2.2. If $r < 1$, as shown, distances decrease; if $r > 1$, distances increase.)

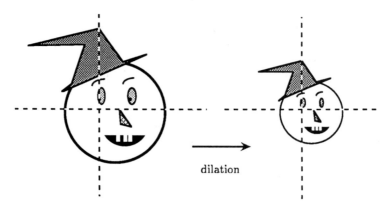

dilation

Figure 1.2.2. A dilation of the plane.

(1.2.3) EXERCISE. *A dilation f maps lines to lines: that is, if L is a line, then the set $f[L] = \{ f(x) : x \in L \}$ is also a line. A dilation f preserves angles:* that is, if lines L_1 and L_2 meet at angle θ, then lines $f[L_1]$ and $f[L_2]$ also meet at angle θ.*

The Sierpiński gasket was constructed above using approximations S_k. Let f_1, f_2, f_3 be the three dilations with ratio $1/2$ and centers at the three vertices of the triangle S_0. Now it follows by induction that

$$S_{k+1} = f_1[S_k] \cup f_2[S_k] \cup f_3[S_k].$$

Then, in much the same way as in Proposition 1.1.9, we can see that

$$S = f_1[S] \cup f_2[S] \cup f_3[S]$$

by showing each side is a subset of the other. So S is the invariant set of the iterated function system (f_1, f_2, f_3).

Coordinates. There is a description of the Sierpiński gasket in terms of coordinates.

(1.2.4) EXERCISE. *Let coordinates (u, v) be defined in the plane with origin at one corner of the triangle S_0, and axes along two of the sides of S_0. Then coordinates (u, v) with $0 \le u \le 1$, $0 \le v \le 1$ represent a point of the Sierpiński gasket if and only if the base 2 expansions of u and v never have 1 in the same place.*

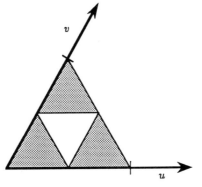

Figure 1.2.5. Coordinate system.

---

*We say that f is conformal.

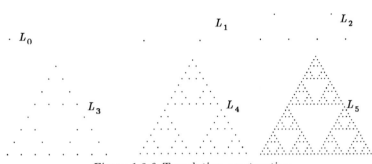

Figure 1.2.6. Translation construction.

Another description of the condition on the base 2 expansions of u and v is to say that the sum $u + v$ can be computed (base 2) without carrying. You will also need to take into account the numbers with two different expansions in base 2.

The preceding exercise may suggest a "translation" type construction for the Sierpiński gasket. Start with a single point. Choose two directions, at a 60 degree angle with each other. Start with a set L_0 containing that one point, and a number $s_0 = 1/2$. The next set is the union of three sets: L_0 and the translates of L_0 through distance s_0 in the two chosen directions. Then let $s_1 = (1/2)s_0$, and let L_2 consist of L_1 together with the translates of L_1 through distance s_1 in the two directions. And so on. (See Figure 1.2.6.)

We say that a plane set L is **dense** in a set S iff $L \subseteq S$ and every point of S is the limit of a sequence of points of L.

(1.2.7) EXERCISE. *Prove that the union $L = \bigcup_{k \in \mathbb{N}} L_k$ is dense in the Sierpiński gasket S.*

Consider Pascal's triangle:

$$
\begin{array}{ccccccccc}
 & & & & 1 & & & & \\
 & & & 1 & & 1 & & & \\
 & & 1 & & 2 & & 1 & & \\
 & 1 & & 3 & & 3 & & 1 & \\
1 & & 4 & & 6 & & 4 & & 1 \\
 & & & & \text{etc.} & & & &
\end{array}
$$

Now, if we make a black dot wherever there is an odd number, and leave blank wherever there is an even number, we will get a geometric arrangement in the plane (Figure 1.2.8).

Figure 1.2.8. Pascal's triangle modulo 2.

(1.2.9) EXERCISE. *Why does the figure look like the Sierpiński gasket?*

1.3. A SPACE OF STRINGS

An **infinite binary tree** is pictured in Figure 1.3.1. It is supposed to continue indefinitely at the top. Each **node** has two nodes immediately above it, called its **children**. (Sometimes it might be convenient to distinguish them as the **left child** and the **right child**.) The node at the bottom, with no parent, is called the **root** of the tree. (Sometimes the tree is drawn the other side up; then it looks less like a "tree", but terminology such as "child" is more reasonable.)

What is a more concrete model of this structure; a model that could be used to investigate the properties of an infinite binary tree?

We consider two symbols, say **0** and **1**. Then we consider finite **strings** (or **words**) made up of these symbols. For example

001010011

The number of symbols in a string α is called the **length** of the string, and written $|\alpha|$. The string above has length 9. How many strings of length n are

etc.

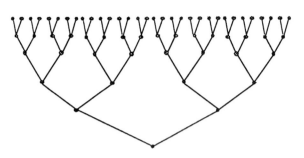

Figure 1.3.1. Infinite binary tree.

there? By convention, we say that there is a unique string of length 0, called the **empty string**, which will be denoted Λ.

If α and β are two strings, then we may form a string $\alpha\beta$, called the **concatenation**, by listing the symbols of the string α followed by the symbols of the string β.

The set of all such finite strings (from the alphabet $E = \{\mathbf{0}, \mathbf{1}\}$) can be identified with the infinite binary tree: The root of the tree corresponds to Λ; if α is a string, then the left child of α is $\alpha\mathbf{0}$ and the right child of α is $\alpha\mathbf{1}$.

We will write $E^{(n)}$ for the set of all strings of length n from the alphabet E. (Recall that $E^{(0)}$ has one element Λ.) We will write

$$E^{(*)} = E^{(0)} \cup E^{(1)} \cup E^{(2)} \cup \cdots$$

for the set of all finite strings. Because of the correspondence with the nodes of the infinite binary tree, we may sometimes refer to $E^{(*)}$ itself as the infinite binary tree. String α represents an ancestor of string β in the tree if and only if α is an **initial segment** of β, that is, $\beta = \alpha\gamma$ for some string γ. Or, α is a **prefix** of β. We will write $\alpha \le \beta$ in this case. If $|\alpha| \ge n$, we write $\alpha{\restriction}n$ for the **initial segment** of α of length n; it consists of the first n symbols of the string α. Thus, $(\mathbf{001010011}){\restriction}4 = \mathbf{0010}$.

Next, let $E^{(\omega)}$ be the set of *infinite* strings from the alphabet E. For $\sigma \in E^{(\omega)}$ there are initial segments $\sigma{\restriction}n$ of all sizes; we can think of an infinite string σ in terms of its initial segments, beginning with the empty string Λ:

$$\Lambda = \sigma{\restriction}0$$
$$< \sigma{\restriction}1$$
$$< \sigma{\restriction}2$$
$$< \cdots .$$

Another way to describe an element $\sigma \in E^{(\omega)}$ is as an infinite sequence of letters from $E = \{0, 1\}$.

Let A be a subset (finite or infinite) of the infinite binary tree $E^{(*)}$. A node $\gamma \in E^{(*)}$ is a **lower bound** for the set A iff $\gamma \leq \alpha$ for all $\alpha \in A$. (When we think of strings, we might say that γ is a "common prefix" for the set A.) A node β is a **greatest lower bound** for the set A iff β is a lower bound for A and $\gamma \leq \beta$ for any other lower bound γ for A. (In string language, β is the "longest common prefix" for the strings in the set A.)

(1.3.2) PROPOSITION. *Every nonempty subset A of the infinite binary tree $E^{(*)}$ has a unique greatest lower bound.*

PROOF: First, choose some $\gamma \in A$. This is possible since A is not empty. Let $n = |\gamma|$ be the length of γ. Consider the integers k with $0 \leq k \leq n$. Some of them, for example 0, have the property that $\gamma \restriction k$ is a lower bound for A. Let k_0 be the largest such k. (A finite nonempty set of integers has a largest element.) I claim that $\gamma \restriction k_0$ is the desired greatest lower bound.

First, $\gamma \restriction k_0$ is a lower bound for A. Let β be any other lower bound for A. Now $\gamma \in A$, so $\beta \leq \gamma$. That means $\beta = \gamma \restriction k$ for some k with $0 \leq k \leq n$. By the definition of k_0, we know that $k \leq k_0$. So $\beta \leq \gamma \restriction k_0$. Thus $\gamma \restriction k_0$ is the greatest lower bound of A.

If α and β are both greatest lower bounds of the set A, then each is $\leq$ the other, so they are equal. ☺

We will be interested in certain subsets of $E^{(\omega)}$. If $\alpha \in E^{(*)}$ is a finite string, let

$$[\alpha] = \left\{ \sigma \in E^{(\omega)} : \alpha \leq \sigma \right\},$$

the set of all infinite strings that begin with α. Now the two children of α are $\alpha 0$ and $\alpha 1$. The corresponding sets satisfy

$$[\alpha] = [\alpha 0] \cup [\alpha 1], \qquad [\alpha 0] \cap [\alpha 1] = \varnothing.$$

We can think of the usual base 2 system as defining a function h from strings $E^{(\omega)}$ to real numbers, by adding a binary point on the left. For example, the periodic string $\sigma = 001001001 \cdots$ corresponds to the real number

$$h(\sigma) = (0.001001001\cdots)_2 = \frac{1}{7}.$$

The set $h\left[E^{(\omega)}\right]$ of values of h is exactly $[0, 1]$.

I have made assertions before about when two strings correspond to the same real number. Can you prove them?

(1.3.3) EXERCISE. *What are necessary and sufficient conditions on infinite strings $\sigma, \tau \in E^{(\omega)}$ so that $h(\sigma) = h(\tau)$?*

A situation like the one just described will be called a "string model": the set of interest, such as $[0, 1]$, is related to a set $E^{(\omega)}$ of strings, called the "model", by a function, such as $h \colon E^{(\omega)} \to \mathbb{R}$, called the "model map". The model can be used to study the set of interest. If the model map h is understood, sometimes we may say that the infinite string σ is the **address** of the point $h(\sigma)$.

Here is a second example of a string model. The set of interest is the triadic Cantor dust C. The space of strings is again the set $E^{(\omega)}$ of infinite strings from the two-letter alphabet $\{\mathbf{0}, \mathbf{1}\}$. But now the map $h \colon E^{(\omega)} \to \mathbb{R}$ is slightly different. Basically what we want to do is to have the letters $\mathbf{0}$ and $\mathbf{1}$ correspond to digits 0 and 2 respectively, and write the numbers in base 3.

So, for example, the periodic string $\sigma = \mathbf{001001001} \cdots$ corresponds to the real number

$$h(\sigma) = (0.002002002 \cdots)_3 = \frac{1}{13}.$$

According to Proposition 1.1.5, the range of h is exactly the Cantor dust: $h\big[E^{(\omega)}\big] = C$. The model map h is related to the two dilations associated with C above:

(1.3.4) EXERCISE. *Let $h \colon E^{(\omega)} \to \mathbb{R}$ be the model map just defined. Then for all strings $\sigma \in E^{(\omega)}$,*

$$h(\mathbf{0}\sigma) = \frac{h(\sigma)}{3}$$

$$h(\mathbf{1}\sigma) = \frac{h(\sigma) + 2}{3}.$$

1.4. TURTLE GRAPHICS

Many of the examples we will discuss are defined recursively. At least the finite approximations to the sets can be drawn by a computer graphics program. When it seems useful, the Logo programs for the sets will be included in the book. In this section, the few elements of the Logo language are discussed. The main part of Logo that we will be concerned with is drawing with "turtle graphics".

Other computer languages could be used in place of Logo. The main requirement is the existence of graphics commands. Abelson and diSessa [1, Appendix B] discuss how to implement turtle graphics in other languages (BASIC, Pascal, APL, Lisp, Smalltalk). For example, in Pascal we would define routines `FORWARD`, `BACK`, `LEFT`, `RIGHT`, and then use the Logo programs with appropriate changes in syntax (parentheses and commas).

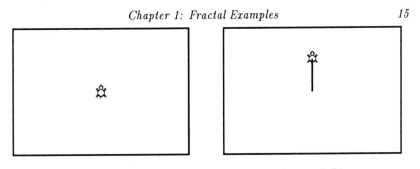

forward 50
Figure 1.4.1a. Turtle   *Figure 1.4.1b.*

Logo. We should think of drawing in the plane, as represented by the computer screen. Our drawing instrument, known as a turtle, is pictured in Figure 1.4.1a. (On some versions of Logo, it may be simplified to a triangle.) Its properties include a "position" (a point in the plane) and a "heading" (the direction the turtle faces).

The commands **forward** and **back** make the turtle move. The argument is the distance to move, measured in some convenient units. As it moves, the turtle draws a line.

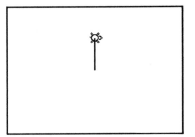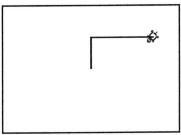

right 90   forward 100 right 135
Figure 1.4.1c.   *Figure 1.4.1d.*

The commands **left** and **right** turn the turtle (change the heading). The angle is measured in degrees.

Between **penup** and **pendown**, no drawing occurs.

The command **repeat** can be used for repetition. Its first argument is the number of times to repeat, and the second argument is the list of commands to be repeated.

Variables can be used to store values. A colon preceding the name of the variable means that we want to refer to the value of the variable. A double-quote preceding the name of the variable means that we want to refer to the name of the variable itself. One of the Logo assignment statements is **make**. To assign a value to the variable, **make** needs to know its name, not its old value.

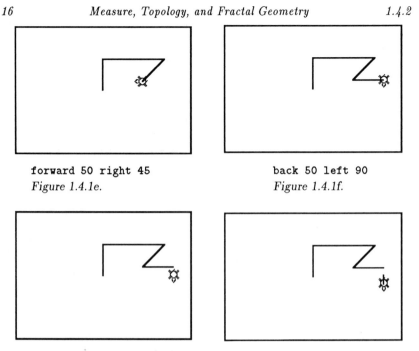

forward 50 right 45   back 50 left 90
Figure 1.4.1e.   *Figure 1.4.1f.*

penup forward 25   pendown forward 25
Figure 1.4.1g.   *Figure 1.4.1h.*

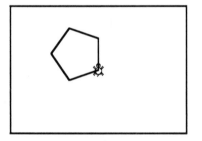

Figure 1.4.2. repeat 5 [forward 50 left 72].

But on the other hand, **repeat** needs to know the value 5, not the name of the variable.

```
make "n 5
repeat :n [forward 50 left 360/:n]
```

The usual arithmetic can be performed: addition :x+:y, subtraction :x-:y, multiplication :x*:y, division :x/:y, square root **sqrt** :x. Some versions of Logo also have powers :x^:y.

Commands can be combined to make new commands. Here is a definition of a command `polygon`. It will have as arguments the length of each side, and the number of sides. The key words `to` and `end` show such a definition.

```
'to polygon :size :n
    repeat :n [forward :size left 360/:n]
end
```

After `polygon` has been defined, it may be used like any other command (Figure 1.4.3). Polygons with more and more sides (of shorter and shorter lengths) converge to a circle. (Convergence for sets is discussed in Sections 2.3 and 24.) Since the graphics screen of the computer (and the ink on this page, as well) has only finite resolution, drawing a regular polygon with enough sides is the same as drawing a circle.

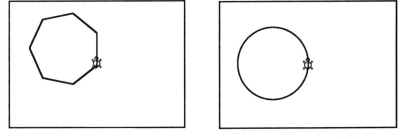

polygon 50 7
Figure 1.4.3a.

polygon 1 360
Figure 1.4.3b.

(1.4.4) EXERCISE. *Write an ellipse program, using only the Logo commands discussed above.*

The first argument of the `if` command is a condition to be tested. If it is true, then the second argument, a list of commands, is executed. The `stop` command ends the execution of the routine. Control returns to wherever the routine was called from. Recursive routines will be used frequently. In the following example, the routine `spiral` calls itself to draw a spiral at a smaller size. The `if` condition insures that it does not run indefinitely.

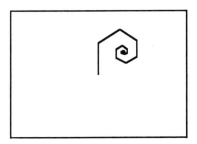

Figure 1.4.5. `hideturtle spiral 50.`

```
to spiral :size
    if :size < 1 [stop]
    forward :size
    right 60
    spiral :size*0.8
end
```

1.5. SETS DEFINED RECURSIVELY

We will consider several sets that are defined recursively. Some of the ways used for defining the triadic Cantor dust and the Sierpiński gasket are recursive. Most of the examples in this section are "dragon" curves. They can be described well using a recursive Logo program.

The Koch curve. The first construction is of "trema" type. (Figure 1.5.1.) It begins with a triangle L_0 (including the interior) having angles of 120, 30, 30 degrees. This triangle can be subdivided into three smaller triangles: two isosceles triangles, angles 120, 30, 30, with long sides along the short sides of the original triangle; and one equilateral triangle. For the next approximation L_1, the trema to be removed is the equilateral triangle. Then we repeat, again and again. The Koch curve is the "limit". (Why is it called a "curve"? See Proposition 2.3.10.)

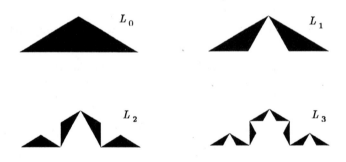

Figure 1.5.1. Koch curve.

A second construction for the Koch curve is of "dragon" type. It involves approximations that are polygons. Here is a program in the Logo language. This is a good example of a "recursive program". The procedure Koch calls itself four times to draw four copies of the curve at 1/3 the size. But each of these four calls of Koch calls four more. And so on. The variable depth is used to limit the recursion to a certain number of levels.

```
to Koch :depth :size
    if :depth = 0 [forward :size stop]
    Koch :depth-1 :size/3
    left 60
    Koch :depth-1 :size/3
    right 120
    Koch :depth-1 :size/3
    left 60
    Koch :depth-1 :size/3
end
```

The first set P_0 is a line segment: its picture (Figure 1.5.2) is obtained by executing Koch 0 200. The value 200 is simply a convenient size. The next approximation P_1 is obtained by executing Koch 1 200. And so on. The set P_k consists of 4^k line segments of length 3^{-k}. The **Koch curve** is the "limit" P of the sequence P_k.

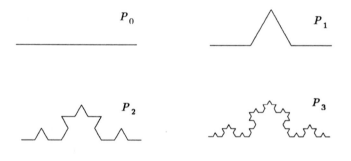

Figure 1.5.2. Results of program Koch.

The Koch curve can be obtained as an iterated function system construction. For example, it is made up of 4 parts, each similar to the whole. See Plate 1. (Similarities are discussed in Section 2.1.)

Three copies of the Koch curve, originating from three sides of an equilateral triangle, form a simple closed curve, often known as the **snowflake curve** (Figure 1.5.3).

Heighway's dragon. (See Figure 1.5.4.) Heighway's dragon is a set in the plane. The approximation P_0 is a line segment of length 1. The next approximation is P_1; it is obtained from P_0 by replacing the line segment by a polygon with two segments, each of length $1/\sqrt{2}$, joined at a right angle. The two ends are the same as before. (There are two choices of how this can be done. We

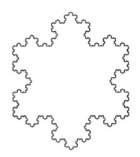

Figure 1.5.3. Snowflake.

choose the one on the "left" side.) For P_2, each line segment in P_1 is replaced by a polygon with two segments, each having length $1/\sqrt{2}$ times the length of the segment that is replaced. The choices alternate between left and right, starting with left, counting from the endpoint on the bottom. **Heighway's dragon** is the "limit" P of this sequence P_n of polygons.

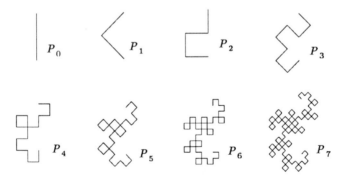

Figure 1.5.4. Heighway's dragon.

The program below is shorter than the description given above, and (with a little study) is less prone to be misinterpreted.* This is a good reason for using Logo to describe the construction.

---

*Did you figure out what I was saying about "left" and "right"?

```
; Heighway's Dragon
make "factor 1/sqrt 2
to heighway :depth :size :parity
    if :depth = 0 [forward :size stop]
    left :parity*45
    heighway :depth-1 :size*:factor 1
    right :parity*90
    heighway :depth-1 :size*:factor (-1)
    left :parity*45
end
```

The third argument is supposed to be either 1 or -1, depending on whether the next corner is supposed to go to the left or to the right. To generate P_0, execute heighway 0 200 1. To generate P_1, execute heighway 1 200 1.

(1.5.5) PROPOSITION. *All of the approximations P_n remain in some bounded region of the plane.*

PROOF: Every point of P_0 has distance at most 1 from the endpoint. Every point of P_1 has distance at most $1/2$ from some point of P_0. We can see by induction, that every point of P_k has distance at most $(1/\sqrt{2})^{k+1}$ from some point of P_{k-1}. Therefore, any point of P_k has distance from the endpoint at most

$$1 + \sum_{j=1}^{k} \left(\frac{1}{\sqrt{2}}\right)^{j+1} < 1 + \sum_{j=1}^{\infty} \left(\frac{1}{\sqrt{2}}\right)^{j+1} .$$

This is a geometric series with ratio < 1, so it converges to a finite value. ☺

The following exercise may be easier to approach after the discussion of similarities (Section 2.1) and convergence of a sequence sets (Sections 2.3 and 2.4).

(1.5.6) EXERCISE. *There are two similarities f_1 and f_2 of the plane onto itself, with ratio $1/\sqrt{2}$, so that Heighway's dragon P satisfies the equation $P = f_1[P] \cup f_2[P]$.*

(1.5.7) PROPOSITION. *In an approximation P_k of Heighway's dragon, the polygon never crosses itself.*

PROOF: It is possible that P_k visits some point twice. But I will show that it does not traverse any line segment twice, and when it visits a point twice, it does not cross itself. The vertices of the polygon P_k lie on a square lattice L_k with edge length $(1/\sqrt{2})^k$. Suppose P_k visits a point twice without visiting an entire line segment twice; then that point must be a vertex of the square lattice.

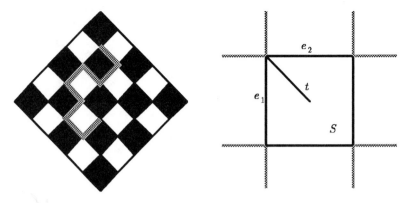

Figure 1.5.7a. Square lattice.   *Figure 1.5.7b*. The square S.

At each vertex, the polygon has a right-angle corner. So it does not cross itself there.

I must only show that P_k does not include a line segment more than once. If it does, it must include a complete edge of the square lattice. Let k be an integer such that P_k does not include any line segment more than once. Let S be some square of the square lattice L_k. I will show that P_{k+1} does not include more than once any of the four line segments of L_{k+1} inside S.

Let t be an edge of L_{k+1} inside S. If t were traversed by P_{k+1} more than once, then the two sides of S adjacent to t (e_1 and e_2 in the figure) must both be traversed by P_k. Now let us color the squares of L_k in a checker-board pattern. The square to the left of the first line segment of P_k will be colored black, and the squares will alternate white and black. Since there is a right-angle turn between two consecutive segments of P_k, the square to the left is always the black square. By convention, let us say that the first line segment is "vertical". The edges alternate between vertical and horizontal. Now when P_{k+1} is constructed, the new pairs of edges are placed alternately to the left and to the right of the old edges. So the new edges corresponding to a vertical edge of P_k will be in the black square bordered by the edge, and the new edges corresponding to a horizontal edge of P_k will be in the white square bordered by the edge. So not both of the edges e_1 and e_2 will produce new edges inside S. ☺

Using an angle other than 90 degrees, we can obtain variant forms of Heighway's dragon. In Figure 1.5.8, the angle is 120 degrees. (For this version, left and right have been reversed). Three copies of this curve, joined in an equilateral triangle, surround a set known as the **fudgeflake**. (See [36, p. 72].) Mandelbrot considered this shape to be derived by "fudging" the snowflake (Figure 1.5.3) using alternating left and right. A fudgeflake is made up of 3 small fudgeflakes; so the fudgeflake tiles the plane (page 36). A set with fractal boundary that tiles the plane is known as a **fractile**.

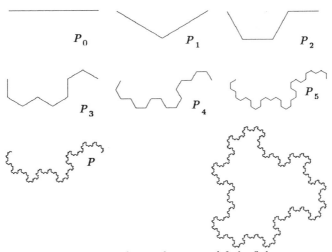

Figure 1.5.8. 120-degree dragon and fudgeflake.

Sierpiński dragon. Here is another dragon. (See Figure 1.5.9.)

```
to SD :depth :size :parity
    if :depth = 0 [forward :size stop]
    left 60*:parity
    SD :depth-1 :size/2 (-:parity)
    right 60*:parity
    SD :depth-1 :size/2 :parity
    right 60*:parity
    SD :depth-1 :size/2 (-:parity)
    left 60*:parity
end
```

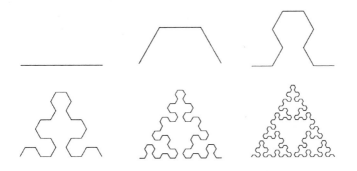

Figure 1.5.9. Sierpiński dragon.

(1.5.10) EXERCISE. *What does this dragon construction have to do with the Sierpiński gasket?*

McWorter's pentigree. Figure 1.5.11 illustrates another dragon; we will call it **McWorter's pentigree**. It is a subset of the plane. The first approximation P_0 is a line segment. In any future stage P_n, each line segment of P_n is replaced by six line segments in a particular pattern to form P_{n+1}. Here is a Logo program. Why was the value $(3 + \sqrt{5})/2$ chosen for the value of the variable `shrink`?

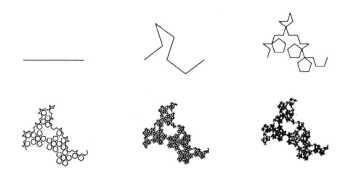

Figure 1.5.11. McWorter's pentigree.

```
; McWorter's pentigree
make "shrink (3 + sqrt 5)/2
to pent :depth :size
    if :depth = 0 [forward :size stop]
    left 36
    pent :depth-1 :size/:shrink
    left 72
    pent :depth-1 :size/:shrink
    right 144
    pent :depth-1 :size/:shrink
    right 72
    pent :depth-1 :size/:shrink
    left 72
    pent :depth-1 :size/:shrink
    left 72
    pent :depth-1 :size/:shrink
    right 36
end
```

(**1.5.12**) EXERCISE. *Does an approximation P_n for McWorter's pentigree ever cross itself?*

Five copies of the pentigree fit together to form a set with five-fold rotational symmetry (Plate 2). This set will also be called "the second form of McWorter's pentigree". It can be thought of as made up of 6 sets similar to the whole, with ratio $2/(3 + \sqrt{5}) = (3 - \sqrt{5})/2$ (Plate 3).

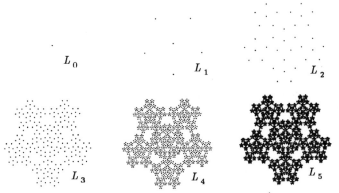

Figure 1.5.13. Translation construction.

Consider the "translation" construction illustrated in Figure 1.5.13. Set L_0 is a single point. Set L_1 is obtained from L_0 by translating it in 5 equally spaced directions by some distance s_0. Set L_2 is obtained from L_1 by translating it in 5 directions (the opposites of the previous directions) by distance $s_1 = rs_0$, where $r = (3 - \sqrt{5})/2$. Set L_3 is obtained by translating L_2 in the original 5 directions by distance $s_2 = rs_1$. And so on.

(**1.5.14**) EXERCISE. *What does this construction have to do with McWorter's pentigree?*

The golden rectangle fractal. The number $\varphi = (1 + \sqrt{5})/2$ is known as the **golden section**. A rectangle with sides in the ratio $1 : \varphi$ is called a **golden rectangle**. It can be decomposed into a square and a rectangle similar to the original rectangle, but reduced by a factor of φ (Figure 1.5.15).

The next example is the **golden rectangle fractal**. It is defined by two Logo routines, as follows:

```
; Golden rectangle fractal, line segment construction.
make "phi (1+sqrt 5)/2
make "phi2 :phi*:phi
make "phi3 :phi2*:phi
make "diag2 (1 + (:phi*:phi))
```

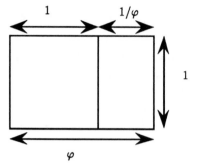

Figure 1.5.15. Golden rectangle.

```
to T :depth :size
    if :depth < 0 [stop]
    forward :size
    penup
        back :size
    pendown
    if :depth = 0 [stop]
    S :depth :size
    penup
        forward (:size*:phi2)/:diag2
        left 90
        back (:size*:phi)/:diag2
    pendown
    T :depth-1 :size/:phi
    penup
        forward (:size*:phi)/:diag2
        right 90
        back (:size*:phi2)/:diag2
    pendown
end
to S :depth :size
    if :depth < 1 [stop]
    penup
        forward :size/:phi2
        right 180
    pendown
    T :depth-2 :size/:phi2
    penup
        forward (:size/:phi2)/:diag2
        right 90
        forward (:size/:phi3)/:diag2
```

```
pendown
T :depth-2 :size/:phi2
penup
    forward (:size/:phi2)/:diag2
    right 90
    forward (:size/:phi3)/:diag2
    forward (:size/:phi2)/:diag2
    right 90
    forward (:size/:phi3)/:diag2
pendown
T :depth-2 :size/:phi2
penup
    forward (:size/:phi2)/:diag2
    right 90
    forward (:size/:phi3)/:diag2
    right 90
pendown
S :depth-3 :size/:phi3
penup
    right 90
    back :size/:phi2
pendown
end
```

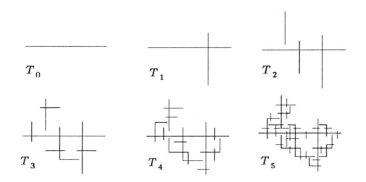

Figure 1.5.16. Golden rectangle fractal.

We may think of two sequences of sets, S_k and T_k. The set S_k is made up of a set similar to S_{k-3} and three sets similar to T_{k-2}. The set T_k is made up of a set isometric to S_k and a set similar to T_{k-1}. The sets S_k should converge to a set S, and the sets T_k should converge to a set T. Then S is made up of a set similar (with ratio $1/\varphi^3$) to S and three sets similar (with ratio $1/\varphi^2$) to T. The set T is made up of a set isometric to S and a set similar (with ratio $1/\varphi$) to T. (See Plate 4.) The "iterated function system" in this case is a bit more complicated than in the previous cases. We will discuss this sort of situation in Chapter 4.

1.6. Number Systems

Let us consider the usual decimal representation of real numbers, and how it can be generalized. Suppose we have a number b for the base (or "radix") of our number system, and a finite set $D = \{d_1, d_2, \cdots, d_k\}$ of numbers, called "digits". We will always assume that 0 is one of the digits. A "whole number" for this system will have the form

$$(1) \qquad \sum_{j=0}^{M} a_j b^j$$

where each $a_j \in D$. Write W for the set of whole numbers. A "fraction" for this system will have the form

$$(2) \qquad \sum_{j=-\infty}^{-1} a_j b^j$$

where each $a_j \in D$. Write F for the set of fractions. The general number represented by this system is the sum of one of each:

$$(3) \qquad \sum_{j=-\infty}^{M} a_j b^j.$$

In order for the representations (2) to converge, we must have $|b| > 1$. Our usual decimal number system has

$$b = 10 \quad \text{and} \quad D = \{0, 1, 2, 3, 4, 5, 6, 7, 8, 9\}.$$

The binary number system has

$$b = 2 \quad \text{and} \quad D = \{0, 1\}.$$

In these two cases, the numbers of the form (1) are exactly the nonnegative integers, and the numbers of the form (2) are the elements of the interval $[0, 1]$. So the numbers of the form (3) are exactly the nonnegative real numbers, $[0, \infty)$.

In both of these cases, there are numbers that have no representation in the form (3), namely the negative numbers.

(1.6.1) EXERCISE. *Let $b = -2$ and $D = \{0,1\}$. The numbers of the form (1) are exactly the integers. The numbers of the form (2) constitute the closed interval $[-2/3, 1/3]$. Every real number has the form (3).*

Both the decimal and binary systems have the fault that some numbers have two different representations. For example

$$0 \cdot 10^{-1} + \sum_{j=2}^{\infty} 9 \cdot 10^{-j} = 1 \cdot 10^{-1} + \sum_{j=2}^{\infty} 0 \cdot 10^{-j}.$$

A system related to the Cantor set avoids this fault.

(1.6.2) EXERCISE. *Let $b = 3$ and $D = \{0,2\}$. Then no number has two different representations in the form (3).*

But in this case, some numbers have no representation at all (such as $1/2$). Can we avoid both problems at once?

(1.6.3) EXERCISE. *Let b be a real number with $|b| > 1$, and let D be a finite set of real numbers including 0. Then either some real number has no expansion in the form (3) or some real number has more than one expansion in the form (3).*

Next, let us consider a number system to represent complex numbers. Now the base b may be a complex number, and the digit-set D is a finite set of complex numbers. We are interested in representing complex numbers in the form

$$(3) \qquad \sum_{j=-\infty}^{M} a_j b^j,$$

where all $a_j \in D$.

(1.6.4) EXERCISE. *Let b be a complex number with $|b| > 1$, and let D be a finite set of complex numbers including 0. Then either some complex number has no expansion in the form (3) or some complex number has more than one expansion in the form (3).*

(1.6.5) EXERCISE. *If every complex number has an expansion of the form (3), then in fact there is a complex number with at least three expansions in the form (3).*

One useful property that a number system for complex numbers might have is that the set W of whole numbers is the set of "algebraic integers" of a number field.* Two examples: The complex numbers of the form $a + ib$, where a and b are integers; they are called the **Gaussian integers**. For a second example, let

$$\omega = \frac{-1 + i\sqrt{3}}{2} = \cos(2\pi/3) + i\sin(2\pi/3).$$

---

*An algebraic integer is a complex number that is a zero of a polynomial such that the coefficients are integers and the coefficient of the highest-degree term is 1. The number field $\mathbb{Q}(i)$ is made up of the complex numbers of the form $a + bi$, where a and b are rational numbers. It turns out that such a number is an algebraic integer if and only if a and b are both integers.

Note that $\omega^3 = 1$, and $\omega^2 = \overline{\omega} = 1/\omega$. The complex numbers of the form $a + b\omega$ (a and b integers) are sometimes known as the **Eisenstein integers.**†

(1.6.6) EXERCISE. *Let $b = -1 + i$ and $D = \{0, 1\}$. Describe the set W of whole numbers.*

Figure 1.6.7. Twindragon.

A construction of the set F of "fractions" for the number system with base $b = -1 + i$ and digit set $D = \{0, 1\}$ is illustrated in Figure 1.6.8. For the first approximation, we consider the set W of "whole numbers", in this case the Gaussian integers; they form a regular square lattice S_0 in the complex plane. The points that are closer to 0 than to any other Gaussian integer form a square. Call this square L_0. Next, consider the numbers representable in our system using at most one place to the right of the radix point. They, too, form a square lattice S_1 in the complex plane, but with shorter sides, and at an angle. The set of points closer to 0 than to any other element of S_1 is a square, and the set of points closer to $(.1)_{-1+i}$ than to any other element of S_1 is also a square. These two squares constitute a set L_1, the next approximation of F. When we take two digits to the right of the radix point, we get a set L_2 made up of four squares. Continuing in this way, we obtain F as the limit of a sequence (L_k) of approximations.

The set obtained in this way is the **twindragon**. Plate 5 suggests that it is made up of two copies of Heighway's dragon. Since every complex number can be represented in this number system, the plane is covered by countably many twindragons, namely the sets $w + F$, one for each Gaussian integer w. The sets $w + F$ overlap only in their boundaries, so this constitutes a "tiling" of the plane. The twindragon (we will see later) has a fractal boundary. Therefore the twindragon is a "fractile".

This set F can be seen from the point of view of an iterated function system. The set F is the union of two parts, namely the set F_0 of all numbers of the form

---

†They are the algebraic integers of the number field $\mathbb{Q}(\sqrt{-3}\,)$.

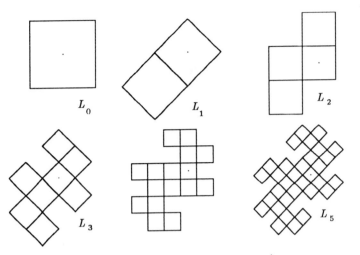

Figure 1.6.8. Twindragon construction.

(2) with $a_{-1} = 0$, and the set F_1 of all numbers of the form (2) with $a_{-1} = 1$. Now the elements of F_0 are exactly b^{-1} times the elements of F; and the elements of F_1 are of the form $b^{-1} + b^{-1}x$, where $x \in F$. So if we write

$$f_0(x) = b^{-1}x,$$
$$f_1(x) = b^{-1} + b^{-1}x,$$

then we have

$$F = f_0[F] \cup f_1[F].$$

So F is the invariant set for this iterated function system. See Plate 6.

(1.6.9) EXERCISE. *Find a complex number with three representations in this system.*

Let us turn next to the "Eisenstein" number system.

(1.6.10) EXERCISE. *Let $b = -2$ and $D = \{0, 1, \omega, \omega^2\}$. Describe the set W of whole numbers.*

For calculations in this system, it may be easier to write $A = \omega$ and $B = \omega^2$. The set F of fractions is pictured in Figure 1.6.11. See also Plate 7.

Figure 1.6.11. Eisenstein fractions.

(**1.6.12**) EXERCISE. *Describe an iterated function system for the Eisenstein fractions.*

The Eisenstein fractions make up another "fractile".

*1.7. REMARKS

Fractal sets, such as those we have seen in this chapter, have been used by mathematicians over the years. Only since Mandelbrot's book has there been interest in them beyond mathematics. Even among mathematicians, these sets had often been considered to have little interest.

I recall a discussion I had with one of my colleagues some years ago. We were talking about a problem dealing with abstract integration. In the course of the discussion, it became apparent that I considered the Cantor set a more natural setting for the problem than the interval. The colleague was surprised by that. He had thought that the Cantor set was only a pathological counterexample. I wonder what he would have said about the Sierpiński gasket.

In 1883, Georg Cantor published a description of the set that today bears his name. It is known as the "Cantor discontinuum", the "Cantor middle-thirds set", or simply the "Cantor set". Over the years is has come to occupy a special place at the heart of descriptive set theory. It was the most important example used by Hausdorff in his paper [**27**] on fractional dimension.

The "iterated function system" was named and popularized by Michael Barnsley [**2**], [**3**].

Wacław Sierpiński published his description of the "Sierpiński gasket" in 1915. This name was assigned to it by Mandelbrot. The pre-Mandelbrot literature calls it something like "Sierpiński's other curve". (It was called a curve, despite its

---

*An optional section.

appearance, because it has topological dimension 1; see Chapter 3. "Sierpiński's curve" was already used to refer to another example, called by Mandelbrot
"Sierpiński's carpet" [36, p. 144].)

The use of "strings" in this book may be a bit unusual. The more conventional
terminology involves "sequences" from the set $\{0, 1\}$, so there is no essential
difference. Many students will be familiar with strings from their computer-
related courses (but probably not infinite strings). Sequences are used in other
ways in this book, so I hope this terminology will reduce the confusion a little.

The description of dragon curves is done naturally using recursive computer
programs. Instead of using a pseudo-code to formulate such programs, or (worse
yet) inventing another ad-hoc notation, I have chosen to use a real computer
language. One of my favorite languages, especially for drawing pictures, is Logo.
Some programmers have negative opinions concerning Logo, because it has been
used to teach small children. But that should not be held against it. There
are programming tasks that I would not use Logo for, but the simple recursive
pictures that are of concern in this book seem suited to Logo. (In fact, almost
all of the pictures in this book were drawn using Logo.)

A reference for turtle graphics is [1]. It contains some interesting ideas on
plane geometry. It contains a solution for Exercise 1.4.4:

```
to ellipse :s  :e
    make "n 1
    repeat 360 [right :n forward :s left :n
    left :n forward (:s * :e) right :n
    make "n :n+1]
end
```

If you show the turtle, and ask him to do `ellipse 1 0.5`, you will see that he
is doing a lot of work.

H. von Koch's curve dates from 1904. It is a continuous curve that has a
tangent line nowhere. The closed "snowflake" version of Figure 1.5.3 is sometimes
used as an example of a curve of infinite length surrounding a finite area. We
will see later that the snowflake curve has Hausdorff dimension strictly larger
than 1. This is a much more precise assertion than merely saying the curve has
infinite length.

Heighway's dragon dates from about 1967; according to Martin Gardner [23],
it was discovered by physicist John E. Heighway, and studied by Heighway together with physicists Bruce A. Banks and William G. Harter. This dragon was
publicized in [11], which contains a wealth of information about the polygonal
approximations. Proposition 1.5.7, which states that a polygonal approximation
does not cross itself, was observed by Harter and Heighway; a proof was published by Davis and Knuth [11]. The proof of Proposition 1.5.7 that appears here
contains elements of a proof submitted by Brian Conrad, Jon Grantham, and
Roger Lee during the Ross summer program. The name "Heighway" is spelled
"Heightway" in some of the references on the subject.

William McWorter described the pentigree in [41]. ("Pentigree" is from "pentagon-filligree".) This book contains the first published analysis of this interesting dragon.

The golden rectangle fractal appears here for the first time. Of course, it was concocted to illustrate [39].

The fractal sets associated with complex number systems are discussed, for example, in [11], [24], [25]. The statement of Exercise 1.6.1 was improved by a proof submitted by Dan Bernstein, Keith Conrad, and Paul Lefelhocz during the Ross summer program.

Here is a hint for Exercise 1.1.2. To see that $1/4$ is not an endpoint, prove by induction that all endpoints are of the form $m/3^k$, for nonnegative integers m, k. Then use unique factorization to show that $1/4$ is not of this form. To show that $1/4 \in C$, prove by induction on k that $1/4 \in C_k$ for all k. In fact, it is easier if you prove more: both $1/4$ and $3/4$ are in C_k for all k.

Si elle était douée de vie,
il ne serait pas possible de l'anéantir,
sans la supprimer d'emblée
car elle renaîtrait sans cesse
des profondeurs de ses triangles,
comme la vie dans l'univers.
—E. Cesàro, 1905

(*If it [the Koch curve] were given life,
it would not be possible to destroy it
except by doing away with it all at once,
since it would be endlessly reborn
of the depths of its triangles,
like life in the universe.*)

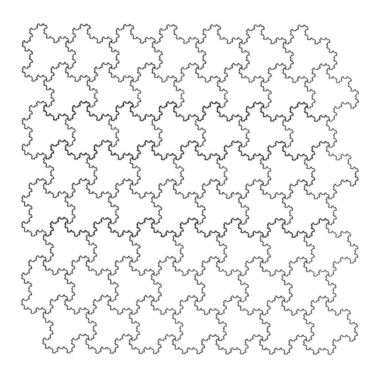

2

Metric Topology

This chapter contains the mathematical background for much of the rest of the book. If the book were organized in strictly logical fashion, then this would be the first chapter of the book; but I included instead some more fractal-like material as Chapter 1. Chapter 2 is a more technical chapter. Have patience! It really is useful for the understanding of the rest of the book.

Mathematics students will eventually learn almost everything in this chapter in the normal course of their studies. So many readers may be able to skip this chapter completely; but it is here for those who need it. Many of the proofs (and exercises) are merely the usual real-number proofs adapted to the setting of metric spaces. So a student who has experience dealing with the proofs of ordinary calculus will see many familiar ideas. A reader who does not care about being mathematically rigorous* could skip the proofs in this chapter. Metric topology is, in fact, important for a lot of modern mathematics. The selection of topics for this chapter was determined by what is required later in the book; so this chapter is a bit peculiar as an introduction to metric topology.

2.1. METRIC SPACES

A **metric space** is a set S together with a function $\rho: S \times S \to [0, \infty)$ satisfying

$$\rho(x, y) = 0 \quad \Longleftrightarrow \quad x = y;$$
$$\rho(x, y) = \rho(y, x);$$
$$\rho(x, z) \leq \rho(x, y) + \rho(y, z).$$

The last inequality is known as the **triangle inequality**: in Euclidean geometry, it says that the sum of the lengths of two sides of a triangle is at least equal to the length of the third side. The nonnegative real number $\rho(x, y)$ is called the **distance** between x and y. The function ρ itself is called a **metric** on the set S. A metric space may be written as a pair (S, ρ), but if the metric is understood, it will be referred to simply as S.

---

*"A simple man believes every word he hears; a clever man understands the need for proof." (Proverbs 14:15, *New English Bible*)

Examples. Let us consider a few examples of metric spaces.

(2.1.1) THEOREM. *The set* $\mathbf{R}$ *of real numbers, with the function*

$$\rho(x, y) = |x - y|$$

is a metric space.

PROOF: First, note that $|x - y| \geq 0$. Also, $|x - y| = 0$ if and only if $x = y$. Next, $\rho(x, y) = |x - y| = |-(x - y)| = |y - x| = \rho(y, x)$. For the triangle inequality, let us consider several cases:

(1) $x \leq y \leq z$: Then $\rho(x, y) + \rho(y, z) = (y - x) + (z - y) = z - x = \rho(x, z)$.
(2) $x \leq z \leq y$: Then $\rho(x, y) + \rho(y, z) = (y - x) + (y - z) \geq y - x \geq z - x = \rho(x, z)$.
(3) $y \leq x \leq z$: Then $\rho(x, y) + \rho(y, z) = (x - y) + (z - y) \geq z - y \geq z - x = \rho(x, z)$.
(4) $y \leq z \leq x$: Then $\rho(x, y) + \rho(y, z) = (x - y) + (z - y) \geq x - y \geq x - a = \rho(x, z)$.
(5) $z \leq y \leq x$: Then $\rho(x, y) + \rho(y, z) = (x - y) + (y - z) = x - z = \rho(x, z)$.
(6) $z \leq x \leq y$: Then $\rho(x, y) + \rho(y, z) = (y - x) + (y - z) \geq y - z \geq x - z = \rho(x, z)$. ☺

(2.1.2) EXERCISE. *When is the triangle inequality actually an equality in the metric space* $\mathbf{R}$?

If d is a positive integer, then $\mathbf{R}^d$ is the set of all ordered d-tuples of real numbers. We can define several operations in this setting. For $x = (x_1, x_2, \ldots, x_d) \in \mathbf{R}^d$, $y = (y_1, y_2, \ldots, y_d) \in \mathbf{R}^d$, and $s \in \mathbf{R}$, define

$$sx = (sx_1, sx_2, \ldots, sx_d);$$
$$x + y = (x_1 + y_1, x_2 + y_2, \ldots, x_d + y_d);$$
$$x - y = x + (-1)y;$$
$$|x| = \sqrt{x_1^2 + x_2^2 + \cdots + x_d^2}.$$

We define d-**dimensional Euclidean space** to be the set $\mathbf{R}^d$ with the metric $\rho(x, y) = |x - y|$.

In order to show that this is a metric space, I will prove two basic inequalities.

(2.1.3) CAUCHY'S INEQUALITY. *Let* $x_1, x_2, \cdots, x_d, y_1, y_2, \cdots, y_d$ *be* $2d$ *real numbers. Then*

$$\left(\sum_{j=1}^{d} x_j y_j \right)^2 \leq \left(\sum_{j=1}^{d} x_j^2 \right) \left(\sum_{j=1}^{d} y_j^2 \right).$$

PROOF: If λ is any real number, then

$$\sum_{j=1}^{d}(x_j - \lambda y_j)^2 \geq 0.$$

Multiplying this out and collecting terms, we see

$$\left(\sum_{j=1}^{d} y_j^2\right)\lambda^2 - 2\left(\sum_{j=1}^{d} x_j y_j\right)\lambda + \left(\sum_{j=1}^{d} x_j^2\right) \geq 0.$$

This is true for all real numbers λ. But in order for a quadratic polynomial $A\lambda^2 + B\lambda + C$ to be nonnegative for all λ, it is necessary that $B^2 - 4AC \leq 0$. In this case, it means

$$4\left(\sum_{j=1}^{d} x_j y_j\right)^2 - 4\left(\sum_{j=1}^{d} y_j^2\right)\left(\sum_{j=1}^{d} x_j^2\right) \leq 0,$$

which is equivalent to the inequality to be proved. ☺

(2.1.4) MINKOWSKI'S INEQUALITY. Let $x, y \in \mathbf{R}^d$. Then $|x + y| \leq |x| + |y|$.

PROOF: Write $x = (x_1, x_2, \cdots, x_d)$ and $y = (y_1, y_2, \cdots, y_d)$. Then

$$|x + y|^2 = \sum_{j=1}^{d}(x_j + y_j)^2$$

$$= \sum_{j=1}^{d} x_j^2 + 2\sum_{j=1}^{d} x_j y_j + \sum_{j=1}^{d} y_j^2$$

$$\leq \sum_{j=1}^{d} x_j^2 + 2\left(\sum_{j=1}^{d} x_j^2\right)^{1/2}\left(\sum_{j=1}^{d} y_j^2\right)^{1/2} + \sum_{j=1}^{d} y_j^2$$

$$= \left(\left(\sum_{j=1}^{d} x_j^2\right)^{1/2} + \left(\sum_{j=1}^{d} y_j^2\right)^{1/2}\right)^2$$

$$= (|x| + |y|)^2.$$

By taking the square root of the extremes, we may conclude (since both of these terms are nonnegative)

$$|x + y| \leq |x| + |y|. \ ☺$$

(2.1.5) COROLLARY. *The space* $\mathbf{R}^d$ *is a metric space with the metric* $\rho(x, y) = |x - y|$.

PROOF: Write $x = (x_1, x_2, \cdots, x_d)$ and $y = (y_1, y_2, \cdots, y_d)$. First,

$$\rho(x, y) = \sqrt{(y_1 - x_1)^2 + (y_2 - x_2)^2 + \cdots + (y_d - x_d)^2} \geq 0.$$

If $\rho(x, y) = 0$, then $(y_1 - x_1)^2 + (y_2 - x_2)^2 + \cdots + (y_d - x_d)^2 = 0$. But a square is nonnegative, so this means that all terms must be 0. That is, $x_j = y_j$ for all j, so that $x = y$. The equation $\rho(x, y) = \rho(y, x)$ is clear. For the triangle inequality, we apply Minkowski's inequality:

$$\rho(x, y) + \rho(y, z) = |x - y| + |y - z|$$
$$\geq |(x - y) + (y - z)| = |x - z| = \rho(x, z). \ \smiley$$

(2.1.6) EXERCISE. *Find necessary and sufficient conditions for Cauchy's inequality to be an equality in* $\mathbf{R}^d$.

(2.1.7) EXERCISE. *Find necessary and sufficient conditions for Minkowski's inequality to be an equality in* $\mathbf{R}^d$.

Next we will consider the set $E^{(\omega)}$ of infinite strings from the two-letter alphabet $E = \{\mathbf{0}, \mathbf{1}\}$. We will define a metric $\rho_{1/2}$ for this space. The basic idea is that two strings should be considered "close" if they begin in the same way.

So let σ and τ be two infinite strings. If $\sigma = \tau$, then of course the distance must be zero:

$$\rho_{1/2}(\sigma, \sigma) = 0.$$

If $\sigma \neq \tau$, then there is a first time they disagree: we can write

$$\sigma = \alpha\sigma'$$
$$\tau = \alpha\tau',$$

where α is a (possibly empty) finite string, and the first character of σ' is different than the first character of τ'. (In the language used before, α is the longest common prefix of σ and τ.) If k is the length of α, then define

$$\rho_{1/2}(\sigma, \tau) = \left(\frac{1}{2}\right)^k.$$

(2.1.8) PROPOSITION. *The set* $E^{(\omega)}$ *is a metric space under the metric* $\rho_{1/2}$.

PROOF: Clearly $\rho_{1/2}(\sigma, \tau) \geq 0$. If $\sigma \neq \tau$, then $\rho_{1/2}(\sigma, \tau) = (1/2)^k > 0$. The equation $\rho_{1/2}(\sigma, \tau) = \rho_{1/2}(\tau, \sigma)$ is also clear.

So all that remains is the triangle inequality. Let σ, τ, θ be three strings. I will prove

$$\rho_{1/2}(\sigma, \tau) \leq \max\{\rho_{1/2}(\sigma, \theta), \rho_{1/2}(\theta, \tau)\}.$$

If two of the strings are equal, then this is clear. So suppose they are all different. Let k be the length of the longest common prefix of σ and θ, and let m be the length of the longest common prefix of θ and τ. If $n = \min\{k, m\}$, we know that the first n letters of σ agree with the first n letters of θ; and the first n letters of θ agree with the first n letters of τ. Therefore, the first n letters of σ agree with the first n letters of τ. So the longest common prefix of σ and τ has length at least n. Therefore:

$$\rho_{1/2}(\sigma, \tau) \leq (1/2)^n = (1/2)^{\min\{k,m\}}$$
$$= \max\{(1/2)^k, (1/2)^m\}$$
$$= \max\{\rho_{1/2}(\sigma, \theta), \rho_{1/2}(\theta, \tau)\}.$$

Finally, this "ultra-triangle" inequality implies the ordinary triangle inequality, since

$$\max\{\rho_{1/2}(\sigma, \theta), \rho_{1/2}(\theta, \tau)\} \leq \rho_{1/2}(\sigma, \theta) + \rho_{1/2}(\theta, \tau). \ \odot$$

Related definitions. If S is a metric space with metric ρ, and $T \subseteq S$, then T is also a metric space with metric ρ_T defined by

$$\rho_T(x, y) = \rho(x, y) \qquad \text{for } x, y \in T.$$

In the future, we will usually write simply ρ rather than ρ_T.

The **diameter** of a subset A of a metric space S is

$$\operatorname{diam} A = \sup\{\rho(x, y) : x, y \in A\}.$$

The diameter of A is the distance between the two most distant points of A, if such points exist. But, for example, if $A = [0, 1)$, the diameter is 1. Even though no two points of A have distance exactly 1, there are pairs x, y of points of A with distance as close as we like to 1; and there are no pairs x, y of points of A with distance greater than 1.

Let S be a metric space, $x \in S$, and $r > 0$. The **open ball** with center x and radius r is the set

$$B_r(x) = \{y \in S : \rho(y, x) < r\}.$$

The **closed ball** with center x and radius r is the set

$$\overline{B_r}(x) = \{y \in S : \rho(y, x) \leq r\}.$$

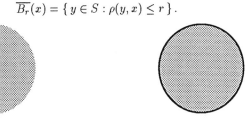

Figure 2.1.9a. Open ball   *Figure 2.1.9b.* Closed ball

Let S be a metric space, and let A be a subset. An **interior point** of A is a point x so that $B_\varepsilon(x) \subseteq A$ for some $\varepsilon > 0$. A set A is called an **open** set iff every point of A is an interior point.

(2.1.10) PROPOSITION. *An open ball $B_r(x)$ is an open set.*

PROOF: Let $y \in B_r(x)$. Then $\rho(x, y) < r$, so that $\varepsilon = r - \rho(x, y)$ is positive. The triangle inequality shows that $B_\varepsilon(y) \subseteq B_r(x)$. So y is an interior point of $B_r(x)$. ☺

(2.1.11) THEOREM. *Let S be a metric space. Then $\varnothing$ and S are open sets. If U and V are open sets, so is $U \cap V$. If $\mathfrak{U}$ is any family of open sets, then the union*

$$\bigcup_{U \in \mathfrak{U}} U$$

is also open.

PROOF: Certainly every point of $\varnothing$ has whatever property I choose to name, such as being an interior point. So $\varnothing$ is an open set.

Let $x \in S$. Then certainly $B_1(x) \subseteq S$. So S is an open set.

Suppose U and V are both open. Let $x \in U \cap V$. Then x is an interior point of U, so there is $\varepsilon_1 > 0$ with $B_{\varepsilon_1}(x) \subseteq U$. Also, x is an interior point of V, so there is $\varepsilon_2 > 0$ with $B_{\varepsilon_2}(x) \subseteq V$. Therefore, if ε is the minimum of ε_1 and ε_2, then we have $B_\varepsilon(x) \subseteq U \cap V$. So $U \cap V$ is an open set.

Let $\mathfrak{U}$ be a family of open sets, and write

$$V = \bigcup_{U \in \mathfrak{U}} U.$$

Let $x \in V$. Then $x \in U$ for some $U \in \mathfrak{U}$. So there is $\varepsilon > 0$ with $B_\varepsilon(x) \subseteq U \subseteq V$. Therefore V is an open set. ☺

Let S be a metric space, and let $A \subseteq S$. A point $x \in S$ is an **accumulation point** of A iff, for every $\varepsilon > 0$, the ball $B_\varepsilon(x)$ contains points of A other than x. A set A is **closed** iff it contains all of its accumulation points. Comparing this to the definition of "open set", we can easily see that a set A is closed if and only if its complement $S \setminus A$ is open.

(2.1.12) EXERCISE. *Let S be a metric space. Then $\varnothing$ and S are closed sets. If A and B are closed sets, so is $A \cup B$. If $\mathcal{C}$ is any family of closed sets, then the intersection*

$$\bigcap_{A \in \mathcal{C}} A$$

is also closed.

(2.1.13) PROPOSITION. *A closed ball $\overline{B_r}(x)$ is a closed set.*

PROOF: Suppose $y \notin \overline{B_r}(x)$. Then $\rho(x, y) > r$, so that $\varepsilon = \rho(x, y) - r$ is positive. The triangle inequality shows that $B_\varepsilon(y) \cap \overline{B_r}(x) = \varnothing$. Therefore y is not an accumulation point of $\overline{B_r}(x)$. This shows that $\overline{B_r}(x)$ is a closed set. ☺

A family $\mathcal{B}$ of open subsets of a metric space S is called a **base for the open sets** of S iff, for every open set $A \subseteq S$, and every $x \in A$, there is $U \in \mathcal{B}$ such that $x \in U \subseteq A$. What would be a good definition for a "base for the closed sets"?

(2.1.14) EXERCISE. *A family $\mathcal{B}$ of open subsets of a metric space S is a base for the open sets if and only if every open set U is of the form*

$$U = \bigcup_{A \in \mathcal{A}} A$$

for some $\mathcal{A} \subseteq \mathcal{B}$.

Of course the definition of "open set" shows that the collection of all open balls is a base for the open sets of S.

An **ultrametric** space S is a metric space for which the metric ρ satisfies the **ultra-triangle inequality**:

$$\rho(x, z) \leq \max\{\rho(x, y), \rho(y, z)\}.$$

Note that in Proposition 2.1.8, we proved that $(E^{(\omega)}, \rho_{1/2})$ is an ultrametric space.

The properties of an ultrametric space may seem strange if you are familiar only with Euclidean space and its subsets. Here are a few examples to help you understand the situation.

(2.1.15) EXERCISE. *Let S be an ultrametric space. Prove:*

(1) *Every triangle is isosceles: that is, if $x, y, z \in S$, then at least two of $\rho(x, y), \rho(y, z), \rho(x, z)$ are equal.*
(2) *A ball $B_r(x)$ of radius r has diameter at most r.*
(3) *Every point of a ball is a center: that is, if $y \in B_r(x)$, then $B_r(x) = B_r(y)$.*
(4) *A closed ball is an open set.*
(5) *An open ball is a closed set.*

Functions on metric spaces. Suppose S and T are metric spaces. A function $h \colon S \to T$ is an **isometry** iff

$$\rho_T(h(x), h(y)) = \rho_S(x, y)$$

for all $x, y \in S$. Two metric spaces are **isometric** iff there is an isometry of one onto the other. A "property" is called a **metric** property iff it is preserved by

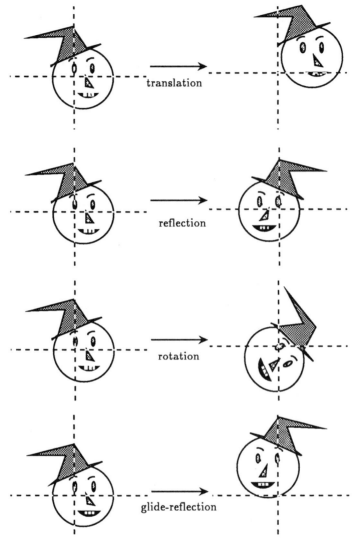

Figure 2.1.16. Isometries of the plane.

isometry, that is: if S and T are isometric, and one has the property, then so does the other.

What are the isometries of the Euclidean plane $\mathbb{R}^2$ into itself? Some examples are pictured in Figure 2.1.16. In fact, the maps of these types are the only isometries of $\mathbf{R}^2$. (But, of course, there are infinitely many isometries of each of the four types.) Here is an outline of the proof:

(1) Let ABC and $A'B'C'$ be two congruent triangles in the plane. Any point P determines a corresponding point P' such that

$$|A - P| = |A' - P'|, |B - P| = |B' - P'|, \text{ and } |C - P| = |C' - P'|.$$

(2) Likewise another point Q yields Q', and $|P - Q| = |P' - Q'|$.

(3) Any two congruent triangles are related by a unique isometry.

(4) Two given congruent line segments AB, $A'B'$ are related by just two isometries: one direct and one opposite.

(5) Any isometry with an invariant point is a rotation or a reflection.

(6) An isometry with no invariant point is a translation or a glide-reflection.

A detailed argument along these lines may be found in [10, Chapter 3].

A function $h\colon S \to T$ is a **similarity** iff there is a positive number r such that

$$\rho(h(x), h(y)) = r\rho(x, y)$$

for all $x, y \in S$. The number r is the **ratio** of h. Two metric spaces are **similar** iff there is a similarity of one onto the other.

(2.1.17) EXERCISE. *Let f be the dilation of $\mathbf{R}^2$ with center a and ratio $r > 0$. Then*

$$|f(x) - f(y)| = r\,|x - y|$$

for all $x, y \in \mathbf{R}^2$.

(2.1.18) EXERCISE. *Describe all of the similarities of two-dimensional Euclidean space $\mathbf{R}^2$ onto itself.*

Let S and T be metric spaces. Let $x \in S$. A function $h\colon S \to T$ is **continuous** at x iff, for every $\varepsilon > 0$, there is $\delta > 0$ such that

$$\rho(x, y) < \delta \quad \Longrightarrow \quad \rho(h(x), h(y)) < \varepsilon.$$

The function h is simply called **continuous** iff it is continuous at every point $x \in S$. This is one of the ten most important definitions in all of mathematics. A thorough understanding of it will be useful to you not only in the study of fractal geometry, but also in much of the other mathematics you will study.

(2.1.19) EXERCISE. *Let $h\colon S \to T$, let $x \in S$, and let $\varepsilon, \delta > 0$. Then we have*

$$\rho(x, y) < \delta \quad \Longrightarrow \quad \rho(h(x), h(y)) < \varepsilon$$

for all $y \in S$ if and only if

$$h[B_\delta(x)] \subseteq B_\varepsilon(h(x)).$$

(2.1.20) EXERCISE. *Isometries and similarities are continuous functions.*

Continuity can be phrased in terms of open sets.

(2.1.21) THEOREM. *A function $h: S \to T$ is continuous if and only if $h^{-1}[V]$ is open in S for all V open in T.*

PROOF: First, suppose that h is continuous. Let V be an open set in T. I must show that $h^{-1}[V]$ is an open set in S. So let $x \in h^{-1}[V]$. Then $h(x) \in V$, which is open, so there is $\varepsilon > 0$ with $B_\varepsilon(h(x)) \subseteq V$. By the continuity of h, there is $\delta > 0$ such that $h[B_\delta(x)] \subseteq B_\varepsilon(h(x)) \subseteq V$. Therefore $B_\delta(x) \subseteq h^{-1}[V]$. So $h^{-1}[V]$ is an open set.

Conversely, suppose that $h^{-1}[V]$ is open in S whenever V is open in T. Let $x \in S$. I must show that h is continuous at x. Let $\varepsilon > 0$. then $B_\varepsilon(h(x))$ is an open set in T. So $W = h^{-1}[B_\varepsilon(h(x))]$ is an open set in S. Now $x \in W$, so there is $\delta > 0$ with $B_\delta(x) \subseteq W$. Therefore $h[B_\delta(x)] \subseteq h[W] \subseteq B_\varepsilon(h(x))$. So h is continuous. ☺

(2.1.22) EXERCISE. *Let $\mathcal{B}$ be a base for the open sets of T. A function $h: S \to T$ is continuous if and only if $h^{-1}[V]$ is open in S for all $V \in \mathcal{B}$.*

A function $h: S \to T$ is a **homeomorphism** of S onto T iff it is bijective, and both h and h^{-1} are continuous. Two metric spaces are **homeomorphic** iff there is a homeomorphism of one onto the other. A "property" is known as a **topological** property iff it is preserved by homeomorphism.

Sequences in a metric space. Let S be a set. A **sequence** in S is, strictly speaking, a function $f: \mathbb{N} \to S$. It is defined by the infinite list of values $f(1), f(2), f(3), \cdots$. Often we will write something like

$$(x_n)_{n \in \mathbb{N}}$$

and understand that the function is specified by $f(1) = x_1$, $f(2) = x_2$, and so on. We may even write simply (x_n).

A sequence (x_n) in a metric space S **converges** to the point $x \in S$ iff for every $\varepsilon > 0$, there is $N \in \mathbb{N}$ so that $\rho(x_n, x) < \varepsilon$ for all $n \geq N$. If this happens, we write $\lim_{n \to \infty} x_n = x$, or $x_n \to x$. Also, x is called the **limit** of the sequence (x_n). We say that the sequence is **convergent** iff it converges to some point.

In order for the definitions to make sense the way they have been worded here, we need to know that limits are unique:

(2.1.23) EXERCISE. *Let (x_n) be a sequence in a metric space S. If $x_n \to a$ and $x_n \to b$, then $a = b$.*

Many of the definitions of this section have equivalent formulations in terms of sequences.

(2.1.24) THEOREM. *Let S and T be metric spaces, and let $h\colon S \to T$ be a function. Then h is continuous if and only if, for every sequence (x_n) in S,*

$$x_n \to x \quad \Longrightarrow \quad h(x_n) \to h(x).$$

PROOF: First, suppose h is continuous. Let (x_n) be a sequence in S, and suppose $x_n \to x$. I must prove that $h(x_n) \to h(x)$. So let $\varepsilon > 0$. Since h is continuous at x, there is $\delta > 0$ with $h[B_\delta(x)] \subseteq B_\varepsilon(h(x))$. Since $x_n \to x$, there is $N \in \mathbb{N}$ so that $x_n \in B_\delta(x)$ for all $n \geq N$. But then $h(x_n) \in B_\varepsilon(h(x))$ for all $n \geq N$. This shows that $h(x_n) \to h(x)$.

For the other direction, I will prove the contrapositive. Suppose h is not continuous. I must prove that the convergence property

$$x_n \to x \quad \Longrightarrow \quad h(x_n) \to h(x)$$

fails. Since h is not continuous, there exist $x \in S$ and $\varepsilon > 0$ such that for all $\delta > 0$, there exists $y \in S$ with $\rho(x, y) < \delta$ but $\rho(h(x), h(y)) \geq \varepsilon$. In particular, for $\delta = 1/n$, there is $x_n \in S$ with $\rho(x_n, x) < 1/n$ but $\rho(h(x_n), h(x)) \geq \varepsilon$. This means that the sequence (x_n) converges to x, but the image sequence $(h(x_n))$ does not converge to $h(x)$. So the convergence property fails. ☺

(2.1.25) EXERCISE. *If $f\colon S_1 \to S_2$ is continuous and $g\colon S_2 \to S_3$ is continuous, then the composition $g \circ f\colon S_1 \to S_3$ is also continuous.*

Let $(x_n)_{n \in \mathbb{N}}$ be a sequence. Suppose we choose an infinite subset of the positive integers, and list them in order:

$$k_1 < k_2 < k_3 < \cdots .$$

Then we may form a new sequence

$$(x_{k_i})_{i \in \mathbb{N}} .$$

This is called a **subsequence** of (x_n).

Let (x_n) be a sequence in a metric space S and let $x \in S$. We say that x is a **cluster point** of the sequence (x_n) iff for every $\varepsilon > 0$, and every $N \in \mathbb{N}$, there exists $n \geq N$ with $\rho(x_n, x) < \varepsilon$.

(2.1.26) PROPOSITION. *The point x is a cluster point of the sequence (x_n) if and only if x is the limit of some subsequence of (x_n).*

PROOF: Suppose x is a cluster point of (x_n). We will define integers $k_1 < k_2 < \cdots$ recursively. Now $1 > 0$, so there is n with $\rho(x_n, x) < 1$. Let k_1 be such an n. Then $1/2 > 0$, so there is $n \geq k_1 + 1$ with $\rho(x_n, x) < 1/2$. Let k_2 be such an n. Suppose k_j has been defined. Then $1/(j+1) > 0$, so there is $n \geq k_j + 1$ with $\rho(x_n, x) < 1/(j+1)$. Let k_{j+1} be such an n. So, we get a sequence $k_1 < k_2 < \cdots$ such that $\rho(x_{k_j}, x) < 1/j$ for all j. Thus x is the limit of the subsequence (x_{k_j}) of (x_n).

Conversely, suppose x is the limit of the subsequence (x_{k_j}) of (x_n). Let $\varepsilon > 0$. Then there is $J \in \mathbb{N}$ so that $\rho(x_{k_j}, x) < \varepsilon$ for $j \geq J$. If $N \in \mathbb{N}$, then there is k_j with both $j \geq J$ and $k_j \geq N$. So we have $\rho(x_{k_j}, x) < \varepsilon$. Therefore x is a cluster point of the sequence (x_n). ☺

(2.1.27) EXERCISE. *Let A be a subset of a metric space S. Then the following are equivalent:*

(1) *A is closed;*
(2) *If $x \in S$ and there is a sequence (x_n) in A such that $x_n \to x$, then $x \in A$.*

(2.1.28) EXERCISE. *Let A be a subset of a metric space S. Then the following are equivalent:*

(1) *A is open;*
(2) *For every $x \in A$ and every sequence (x_n) in S such that $x_n \to x$, there exists $N \in \mathbb{N}$ so that $x_n \in A$ for all $n \geq N$.*

Completeness. A **Cauchy sequence** in a metric space S is a sequence (x_n) satisfying: for every $\varepsilon > 0$ there is $N \in \mathbb{N}$ so that $\rho(x_n, x_m) < \varepsilon$ for all n, m with $n \geq N$ and $m \geq N$. This is an important definition for the real line $\mathbb{R}$. A sequence in $\mathbb{R}$ is convergent if and only if it is a Cauchy sequence. For a general metric space, only one direction remains true in general:

(2.1.29) PROPOSITION. *Every convergent sequence is a Cauchy sequence.*

PROOF: Suppose $x_n \to x$. I will show that (x_n) is a Cauchy sequence. Let $\varepsilon > 0$. Then also $\varepsilon/2 > 0$. Since $x_n \to x$, there is $N \in \mathbb{N}$ such that $\rho(x_n, x) < \varepsilon/2$ for all $n \geq N$. Then, if $n, m \geq N$, we have

$$\rho(x_n, x_m) \leq \rho(x_n, x) + \rho(x, x_m) < \frac{\varepsilon}{2} + \frac{\varepsilon}{2} = \varepsilon.$$

Therefore (x_n) is a Cauchy sequence. ☺

Consider the metric space S consisting of all the rational numbers. The number $\sqrt{2}$ is irrational. But if its decimal expansion is truncated after n places, the result is a rational number:

$$x_1 = 1.4$$
$$x_2 = 1.41$$
$$x_3 = 1.414$$
$$x_4 = 1.4142$$

and so on. This is a Cauchy sequence in the metric space S that does not converge in S. The theorem of Cauchy is so useful, however, that we will single out those metric spaces where it is true: A metric space S is called **complete** iff every Cauchy sequence in S converges (in S).

(2.1.30) EXERCISE. *Three-dimensional Euclidean space* $\mathbb{R}^3$ *is complete.*

(2.1.31) EXERCISE. *Suppose S is an ultrametric space. Then a sequence (x_n) in S is a Cauchy sequence if and only if $\rho(x_n, x_{n+1}) \to 0$.*

(2.1.32) PROPOSITION. *The space $E^{(\omega)}$ of infinite strings from the alphabet $\{0, 1\}$ is complete under the metric $\rho_{1/2}$.*

PROOF: Let (σ_n) be a Cauchy sequence in $E^{(\omega)}$. We first define a "candidate" τ for the limit, then we prove that it is, in fact, the limit. For each k, there is $n_k \in \mathbb{N}$ so that for all $n, m \geq n_k$, we have

$$\rho_{1/2}(\sigma_n, \sigma_m) < \left(\frac{1}{2}\right)^k.$$

That means that $\sigma_{n_k} {\restriction} k = \sigma_m {\restriction} k$ for all $m \geq n_k$. Define τ as follows: The kth letter of τ is the kth letter of σ_{n_k}. So τ satisfies

$$\tau {\restriction} k = \sigma_{n_k} {\restriction} k$$

for all k.

To see that $\sigma_n \to \tau$, let $\varepsilon > 0$ be given. Choose k so that $(1/2)^k < \varepsilon$. Then for $m > n_k$, we have $\sigma_m {\restriction} k = \tau {\restriction} k$, so $\rho_{1/2}(\sigma_m, \tau) \leq (1/2)^k < \varepsilon$. This shows that $\sigma_n \to \tau$. ☺

(2.1.33) EXERCISE. *Completeness is a metric property, but not a topological property.*

The **closure** of a set A is the set $\overline{A}$, consisting of A together with all of its accumulation points. It is a closed set. A set A is **dense** in a set B iff $\overline{A} = B$. It may be useful to verify that this agrees with the definitions given above for $\mathbb{R}$ (page 6) and $\mathbb{R}^2$ (page 10).

A point x that belongs both to the closure of the set A and to the closure of the complementary set $S \setminus A$ is called a **boundary point** of A. The **boundary** of A is the set of all boundary points of A. We will write it as ∂A.

A set has empty boundary if and only if it is clopen.* The boundary of the interval (a, b) in the metric space $\mathbb{R}$ is the two-point set $\{a, b\}$. The boundary of the ball $B_r(x)$ in $\mathbb{R}^d$ is the sphere $\{ y \in \mathbb{R}^d : |x - y| = r \}$. In a general metric space S, however, we can only say that the boundary of the ball $B_r(x)$ is a subset of the sphere $\{ y \in \mathbb{R}^d : \rho(x, y) = r \}$. For example, if S is ultrametric, then this boundary of $B_r(x)$ is empty, even if the sphere is not.

(2.1.34) EXERCISE. *If A is any subset of a metric space, then ∂A is a closed set.*

Suppose $T \subseteq S$. Then T is a metric space in its own right. If $A \subseteq T$, then also $A \subseteq S$. The boundary of A is a concept that depends on the complement of A, as well as on A itself, so it makes a difference whether the boundary is taken in T or in S. For example, $S = \mathbb{R}$ contains $T = [0, 1]$. If $A = [0, 1/2]$, then $\partial_T A = \{1/2\}$, but $\partial_S A = \{0, 1/2\}$.

---

*"Clopen" means "closed and open".

(2.1.35) THEOREM. *Let S be a metric space, $A \subseteq S$, and $T \subseteq S$. Then $\partial_T(A \cap T) \subseteq \partial_S A$.*

PROOF: Let $x \in \partial_T(A \cap T)$. For every $\varepsilon > 0$, there exist points $y \in A \cap T$ with $\rho(y, x) < \varepsilon$ and points $z \in T \setminus (A \cap T)$ with $\rho(z, x) < \varepsilon$. But points such as y are points of A, and points such as z are points of $S \setminus A$. So x is a boundary point of A in S. ☺

Contraction mapping. A point x is a **fixed point** of a function f iff $f(x) = x$. A function $f \colon S \to S$ is a **contraction** iff there is a constant $r < 1$ such that

$$\rho(f(x), f(y)) \le r\rho(x, y)$$

for all $x, y \in S$. A contraction is easily seen to be continuous. There is a useful theorem on contraction mappings.

(2.1.36) CONTRACTION MAPPING THEOREM. *A contraction mapping f on a complete nonempty metric space S has a unique fixed point.*

PROOF: First, there is at most one fixed point. If x and y are both fixed points, then $\rho(x, y) = \rho(f(x), f(y)) \le r\rho(x, y)$. But $0 \le r < 1$, so this is impossible if $\rho(x, y) > 0$. Therefore $\rho(x, y) = 0$, so $x = y$.

Now let x_0 be any point of S. (Recall that S is nonempty.) Then define recursively

$$x_{n+1} = f(x_n) \qquad \text{for } n \ge 0.$$

I claim that (x_n) is a Cauchy sequence. Write $a = \rho(x_0, x_1)$. It follows by induction that $\rho(x_{n+1}, x_n) \le ar^n$. But then, if $m < n$, we have

$$
\begin{aligned}
\rho(x_m, x_n) &\le \sum_{j=m}^{n-1} \rho(x_{j+1}, x_j) \\
&\le \sum_{j=m}^{n-1} ar^j \\
&= \frac{ar^m - ar^n}{1 - r} \\
&= \frac{ar^m(1 - r^{n-m})}{1 - r} \\
&\le \frac{ar^m}{1 - r}.
\end{aligned}
$$

Therefore, if $\varepsilon > 0$ is given, choose N large enough that $ar^N/(1 - r) < \varepsilon$. Then, for $n, m \ge N$, we have $\rho(x_m, x_n) < \varepsilon$.

Now S is complete and (x_n) is a Cauchy sequence, so it converges. Let x be the limit. Now f is continuous, so from $x_n \to x$ follows also $f(x_n) \to f(x)$. But $f(x_n) = x_{n+1}$, so $f(x_n) \to x$. Therefore the two limits are equal, $x = f(x)$, so x is a fixed point. ☺

The preceding theorem can be used to prove the existence of certain points in a complete metric space. But more than that is true: the proof of the theorem shows a way to "construct" the point in question. We record this consequence of the proof.

(2.1.37) COROLLARY. *Let f be a contraction mapping on a complete metric space S. If x_0 is any point of S, and*

$$x_{n+1} = f(x_n) \qquad \text{for } n \geq 0,$$

then the sequence x_n converges to the fixed point of f.

Figure 2.1.37. Illustration for Corollary 2.1.37. "BONZO Dog Food, said the white letters on the orange label, and below the name was a picture of a little black-and-white spotted dog, walking on his hind legs and wearing a chef's cap and an apron. The dog carried a tray on which there was another can of BONZO Dog Food, on the label of which another little black-and-white spotted dog, exactly the same but much smaller, was walking on his hind legs and carrying a tray on which there was another can of BONZO Dog Food, on the label of which another little black-and-white spotted dog, exactly the same but much smaller, was walking on his hind legs and carrying a tray on which there was another can of BONZO Dog Food, and so on until the dogs became too small for the eye to follow."†

A function $f \colon S \to T$ has **bounded increase** iff there is a constant B with

$$\rho(f(x), f(y)) \leq B\,\rho(x, y) \qquad \text{for all } x, y \in S.$$

(Such a function f is also called a **Lipschitz function**.)

(2.1.38) EXERCISE. *Suppose f has bounded increase. Does it follow that f is continuous?*

A function $f \colon S \to T$ has **bounded decrease** iff there is a constant $A > 0$ with

$$\rho(f(x), f(y)) \geq A\,\rho(x, y) \qquad \text{for all } x, y \in S.$$

---

†From *The Mouse and His Child*, by Russell Hoban (Harper & Row, 1967). Illustration © 1967 by Lillian Hoban.

(2.1.39) EXERCISE. *Suppose f has bounded decrease. Does it follow that f is continuous?*

A function $f\colon S \to T$ has **bounded distortion** iff it has both bounded increase and bounded decrease, that is, there are positive constants A, B such that

$$A\,\rho(x, y) \leq \rho(f(x), f(y)) \leq B\,\rho(x, y).$$

Such a function f is also called a **metric equivalence** or **lipeomorphism**.

(2.1.40) EXERCISE. *Suppose f has bounded distortion. Does it follow that f is a homeomorphism?*

(2.1.41) EXERCISE. *Find an example of a sequence F_n of closed sets in $\mathbb{R}$ such that $\bigcup_{n \in \mathbb{N}} F_n$ is not closed.*

(2.1.42) EXERCISE. *Suppose that, for each $n \in \mathbb{Z}$, we have a closed set $F_n \subseteq [n, n+1]$. Then show that $\bigcup_{n \in \mathbb{Z}} F_n$ is closed.*

2.2. SEPARABLE AND COMPACT SPACES

Sometimes it is best to deal not with the most general metric space, but a more specialized class of metric spaces. This section deals with two important special classes of metric spaces.

Separable. A family $\mathcal{U}$ of subsets of S is said to **cover** a set A iff A is contained in the union of the family $\mathcal{U}$. A family which covers a set is known as a **cover** of the set. A cover consisting of a finite number of sets is called a **finite cover**. A cover consisting of a countable number of sets is called a **countable cover**. An **open cover** of a set A is a cover of A consisting only of open sets. If $\mathcal{U}$ is a cover of A, then a **subcover** is a subfamily of $\mathcal{U}$ that still covers A.

(2.2.1) THEOREM. *Let S be a metric space. The following are equivalent:*

   (1) *There is a countable set D dense in S. [S is a **separable** space.]*

   (2) *There is a countable base for the open sets of S. [S satisfies the **second axiom of countability**.]*

   (3) *Every open cover of S has a countable subcover. [S has the **Lindelöf property**.]*

PROOF: (1) $\Longrightarrow$ (2). Suppose S admits a countable dense set D. Let $\mathcal{B} = \{ B_{1/n}(a) : a \in D, n \in \mathbb{N} \}$. Then $\mathcal{B}$ is a countable family of open sets. I claim it is a base of the open sets of S. Let U be any open set of S, and let $x \in U$. Then there is $\varepsilon > 0$ so that $B_\varepsilon(x) \subseteq U$. Choose n so that $2/n < \varepsilon$. Since D is dense in S, we know that $x \in \overline{D}$. So there is a point $a \in B_{1/n}(x) \cap D$. Then $B_{1/n}(a) \in \mathcal{B}$, and we have $x \in B_{1/n}(a) \subseteq B_{2/n}(x) \subseteq U$. Thus $\mathcal{B}$ is a countable base for the open sets of S.

   (2) $\Longrightarrow$ (3). Suppose there is a countable base $\mathcal{B}$ for the open sets of S. Let $\mathcal{U}$ be an open cover of S. For each point $x \in S$, choose a set $U_x \in \mathcal{U}$ with $x \in U_x$.

Then choose a basic set $D_x \in \mathcal{B}$ with $x \in D_x \subseteq U_x$. Now $\{ D_x : x \in S \}$ is a subfamily of $\mathcal{B}$, so it is countable. So it has the form

$$\{ D_x : x \in S \} = \{ D_{x_n} : n \in \mathbb{N} \}.$$

Now write $\mathcal{V} = \{ U_{x_n} : n \in \mathbb{N} \}$. This is a countable subfamily of $\mathcal{U}$. If $x \in S$, then $D_x = D_{x_n}$ for some n, and therefore $x \in D_{x_n} \subseteq U_{x_n}$. So $\mathcal{V}$ is a countable subcover.

(3) $\Longrightarrow$ (1). Suppose that S has the property of Lindelöf. For $n \in \mathbb{N}$, the collection

$$\mathcal{B}_n = \{ B_{1/n}(x) : x \in S \}$$

is an open cover of S. Therefore it has a countable subcover, say

$$\mathcal{A}_n = \{ B_{1/n}(y) : y \in Y_n \},$$

for a countable set Y_n. Let

$$D = \bigcup_{n \in \mathbb{N}} Y_n.$$

Then D is countable, since it is a countable union of countable sets. If $x \in S$ is any point, and $\varepsilon > 0$, choose n with $1/n < \varepsilon$. Since $\mathcal{A}_n$ is a cover of S, there is $y \in Y_n \subseteq D$ with $x \in B_{1/n}(y)$. Therefore $y \in B_{1/n}(x) \subseteq B_\varepsilon(x)$. This shows that D is dense in S. ☺

A metric space S will be called **separable** iff it has one (and therefore all) of the properties of Theorem 2.2.1. A subset of a metric space will be called **separable** iff it is a separable metric space when considered to be a metric space in its own right. Separable metric spaces have many useful properties. By (2), any subset of a separable metric space is separable. By (1), a union of countably many separable sets is separable. All of the examples of metric spaces in this book are separable:

(2.2.2) EXERCISE. *The set*

$$\{ (x_1, x_2, \cdots, x_d) \in \mathbb{R}^d : all \ x_j \ are \ rational \ numbers \}$$

is countable and dense in Euclidean space $\mathbb{R}^d$.

(2.2.3) EXERCISE. *Consider the space* $E^{(\omega)}$ *of infinite strings from the alphabet* $\{0, 1\}$ *under the metric* $\rho_{1/2}$. *The set*

$$\left\{ [\alpha] : \alpha \in E^{(*)} \right\}$$

is a countable base for the open sets. In fact, every open ball is one of the sets $[\alpha]$.

Compact. We will begin by considering compactness of a closed interval $[a, b]$ in $\mathbb{R}$.

(2.2.4) THE BOLZANO-WEIERSTRASS THEOREM. *Let $a < b$ be real numbers. If (x_n) is any sequence in the interval $[a, b]$, then (x_n) has at least one cluster point.*

PROOF: We will define recursively a sequence (I_k) of closed intervals, such that each I_k contains x_n for infinitely many n. Let $I_0 = [a, b]$. Then I_0 contains x_n for all n. If I_k has been defined, say $I_k = [a_k, b_k]$, then we will consider the midpoint $c_k = (a_k + b_k)/2$. Since I_k contains x_n for infinitely many n, either the left half $[a_k, c_k]$ or the right half $[c_k, b_k]$ also contains x_n for infinitely many n. Let I_{k+1} be a half that contains x_n for infinitely many n. Now also note from the definition that the length of I_{k+1} is half the length of I_k, so the length of I_k is $2^{-k}(b - a)$. This converges to 0. Also note that $I_{k+1} \subseteq I_k$. This means that $a_m \in I_k$ for all $m \geq k$. So (a_k) is a Cauchy sequence, and hence a convergent sequence. Let x be its limit. The interval I_k is closed, so the limit x is in I_k. I claim that x is a cluster point of (x_n). If $\varepsilon > 0$, choose k so that the length of I_k is less than ε. Now x_n is in I_k for infinitely many n, so if N is given, then there is $n \geq N$ with $x_n \in I_k$. Also $x \in I_k$, so $|x_n - x| < \varepsilon$. Thus (x_n) has a cluster point. ☺

A metric space S is called **sequentially compact** iff every sequence in S has at least one cluster point (in S).

Let $r > 0$. A subset A of a metric space S is an *r*-net for S iff every point of S is within distance at most r of some element of A. For example, the countable set $\{ rn : n \in \mathbb{Z} \}$ is an *r*-net in $\mathbb{R}$.

(2.2.5) PROPOSITION. *Let S be a sequentially compact metric space, and let $r > 0$. Then S has a finite *r*-net.*

PROOF: Suppose S has no finite *r*-net.

We will define a sequence (x_n) recursively, with $\rho(x_n, x_m) > r$ for all $m \neq n$. First, $S \neq \varnothing$ (since $\varnothing$ is a finite *r*-net in $\varnothing$). So we may choose $x_1 \in S$. Now, assume $x_1, x_2, \cdots, x_n$ have been chosen. Since $\{x_1, x_2, \cdots, x_n\}$ is not an *r*-net, there exists a point (call it x_{n+1}) such that $\rho(x_j, x_{n+1}) > r$ for $1 \leq j \leq n$. This completes the definition of the sequence (x_n).

Now I claim that this sequence (x_n) has no cluster point. If x were a cluster point, then the ball $B_{r/2}(x)$ would contain at least two of the points x_n, which is impossible since they have distance exceeding r. Therefore S is not sequentially compact. ☺

(2.2.6) COROLLARY. *A sequentially compact metric space is separable.*

PROOF: Suppose S is sequentially compact. For each n let D_n be a finite $1/n$-net for S. Then $D = \bigcup_{n \in \mathbb{N}} D_n$ is a countable set dense in S. ☺

(2.2.7) PROPOSITION. *Let $a < b$ be real numbers. If A is any infinite subset of the interval $[a, b]$, then A has at least one accumulation point.*

PROOF: If A is an infinite set, we may choose a sequence $x_n \in A$ of distinct elements. By Theorem 2.2.4, (x_n) has a cluster point x. If $\varepsilon > 0$, then $x_n \in B_\varepsilon(x)$ for infinitely many n, so $x_n \in B_\varepsilon(x)$ for some $x_n \neq x$. This shows that x is an accumulation point of A. ☺

A metric space S is called **countably compact** iff every infinite subset of S has at least one accumulation point (in S).

Let $\mathcal{F}$ be a family of subsets of a set S. We say that $\mathcal{F}$ has the **finite intersection property** iff any intersection of finitely many sets from $\mathcal{F}$ is nonempty.

(2.2.8) THE HEINE-BOREL THEOREM. *Let $a < b$ be real numbers. Let $\mathcal{F}$ be a family of closed subsets of the interval $[a, b]$. If $\mathcal{F}$ has the finite intersection property, then the intersection*

$$\bigcap_{F \in \mathcal{F}} F$$

of the entire family is not empty.

PROOF: First, by Exercise 2.2.2, the line $\mathbb{R}$ is separable; therefore also $[a, b]$ is separable. Suppose (for purposes of contradiction), that

$$(2.2.8a) \qquad \bigcap_{F \in \mathcal{F}} F = \varnothing.$$

That means that $\{ [a, b] \setminus F : F \in \mathcal{F} \}$ is an open cover of $[a, b]$. So by the Lindelöf property (Theorem 2.2.1), there is a countable subcover. That means that there is a countable number of the sets of $\mathcal{F}$ with empty intersection. Say

$$(2.2.8b) \qquad \bigcap_{n \in \mathbb{N}} F_n = \varnothing,$$

where $F_n \in \mathcal{F}$. Now, for each n, the finite intersection $F_1 \cap F_2 \cap \cdots \cap F_n$ is nonempty; choose an element x_n. By sequential compactness (Theorem 2.2.4), the sequence x_n has a cluster point, say x. Since F_n is closed and $x_m \in F_n$ for all $m \geq n$, we have $x \in F_n$. This is true for all n, so

$$x \in \bigcap_{n \in \mathbb{N}} F_n,$$

which contradicts (2.2.8b). This contradiction arose from supposing (2.2.8a). Therefore

$$\bigcap_{F \in \mathcal{F}} F \neq \varnothing. ☺$$

A metric space S is called **bicompact** iff every family of closed sets with the finite intersection property has nonempty intersection. As in the proof above, simply taking complements shows that this is equivalent to saying that every open cover of S has a finite subcover.

(2.2.9) THEOREM. *Let S be a metric space. The following are equivalent:*

(1) *S is sequentially compact,*

(2) *S is countably compact,*

(3) *S is bicompact.*

PROOF: (3) $\Longrightarrow$ (2). Suppose S is not countably compact. Then there is an infinite subset A of S with no accumulation points. For each point $x \in S$, choose an open ball B_x such that B_x contains no points of A (except possibly x itself). Then $\mathcal{U} = \{ B_x : x \in S \}$ is an open cover of S. Any finite subcollection of $\mathcal{U}$ contains only finitely many points of A, so $\mathcal{U}$ does not admit a finite subcover. So S is not bicompact.

(2) $\Longrightarrow$ (1). Suppose S is countably compact. Let (x_n) be any sequence in S. If there is a point x with $x_n = x$ for infinitely many n, then that x is a cluster point of the sequence (x_n). On the other hand, if there is no such point, then the set $A = \{ x_n : n \in \mathbb{N} \}$ of values is an infinite set; so A has an accumulation point, which is easily seen to be a cluster point of the sequence (x_n). So in all cases (x_n) has a cluster point. Thus S is sequentially compact.

(1) $\Longrightarrow$ (3). Suppose S is sequentially compact. Then by Corollary 2.2.6, S is separable. The rest of the proof follows the proof of Theorem 2.2.8 word for word. ☺

A metric space S will be called **compact** iff it has one (and therefore all) of the properties of Theorem 2.2.9. A subset of a metric space will be called **compact** iff it is a compact metric space when considered to be a metric space in its own right. One of the most useful ways to prove that a set is compact is the following:

(2.2.10) PROPOSITION. *A closed subset of a compact space is compact.*

PROOF: Suppose S is compact and $T \subseteq S$ is closed. Let (x_n) be a sequence in T. Now by the compactness of S, there is $x \in S$ which is the limit of a subsequence (x_{k_i}) of (x_n). But T is closed and $x_{k_i} \in T$ for all i, so the limit x is also in T. Thus (x_n) has a cluster point in T. This shows that T is compact. ☺

(2.2.11) PROPOSITION. *Let $A \subseteq \mathbb{R}^d$. Then A is compact if and only if A is closed and bounded.*

PROOF: First, suppose A is closed and bounded. Then A is a subset of a large cube,

$$C = \{ x = (x_1, x_2, \cdots, x_d) : -a \leq x_j \leq a \text{ for all } j \} .$$

By Proposition 2.2.10, if I show that C is compact, it will follow that A is compact. Now let (y_n) be a sequence in C. For notation, write

$$y_n = (y_{n1}, y_{n2}, \cdots, y_{nd}).$$

Now the sequence of first coordinates $(y_{n1})_{n\in\mathbb{N}}$ is a sequence in $[-a, a]$, which is compact. So there is a subsequence that converges, so there is an infinite set $N_1 = \{n_1 < n_2 < \cdots\}$ and

$$\lim_{n\in N_1} y_{n1} \to z_1.$$

Next, the sequence of second coordinates $(y_{n2})_{n\in N_1}$ is a sequence in $[-a, a]$, which is compact. So there is a subsequence that converges, say $N_2 \subseteq N_1$ and

$$\lim_{n\in N_2} y_{n2} \to z_2.$$

Similarly, we get $N_3 \supseteq \cdots \supseteq N_d$, with

$$\lim_{n\in N_j} y_{nj} \to z_j.$$

Finally, the subsequence $(y_n)_{n\in N_d}$ has all coordinates convergent, and its limit is $z = (z_1, z_2, \cdots, z_d) \in C$. This proves that C is compact.

Conversely, suppose that A is compact. If A is unbounded, then

$$\{ B_n(0) \cap A : n \in \mathbb{N} \}$$

is an open cover of A with no finite subcover. If A is not closed, there is an accumulation point x of A that is not in A. So there is a sequence (x_n) in A converging to x. This sequence has no cluster point in A. ☺

(2.2.12) EXERCISE. *The metric space $E^{(\omega)}$ of all infinite strings constructed using the alphabet $E = \{0, 1\}$ is compact under the metric $\rho_{1/2}$.*

Of course compactness is a topological property: If S and T are homeomorphic, and S is compact, then T is compact.

(2.2.13) EXERCISE. *A compact subset of a metric space is closed.*

(2.2.14) EXERCISE. *The union of finitely many compact sets is compact.*

Image and inverse image. If $f\colon S \to T$ is a continuous function, and $A \subseteq S$, some properties of the set A are related to properties of the image set

$$f[A] = \{ f(x) : x \in A \}.$$

(2.2.15) THEOREM. *Let $f\colon S \to T$ be continuous. Let $A \subseteq S$ be compact. Then $f[A]$ is compact.*

PROOF: Let (y_n) be a sequence in $f[A]$. Then there exist points $x_n \in A$ with $f(x_n) = y_n$. By the compactness of A, there is a subsequence (x_{k_i}) that converges to some point $x \in A$. But since f in continuous, this implies that $y_{k_i} = f(x_{k_i}) \to f(x)$. Now $f(x) \in f[A]$, so (y_n) has a cluster point in $f[A]$. This shows that $f[A]$ is compact. ☺

(2.2.16) COROLLARY. *Let S be a compact metric space, and let the function $f\colon S \to \mathbf{R}$ be continuous. Then f is bounded; that is: there is $B \in \mathbf{R}$ such that $|f(x)| \leq B$ for all $x \in S$.*

PROOF: By Theorem 2.2.14, $f[S]$ is a compact subset of $\mathbf{R}$. By Proposition 2.2.11, it is a bounded set. ☺

Let $f\colon S \to T$ be continuous. Properties of a set $B \subseteq T$ may be related to properties of the inverse image

$$f^{-1}[B] = \{\, x \in S : f(x) \in B \,\}.$$

(2.2.17) PROPOSITION. *Let $f\colon S \to T$ be continuous. Consider a set $B \subseteq T$. Then:*

    (1) *If B is open, then $f^{-1}[B]$ is open.*
    (2) *If B is closed, then $f^{-1}[B]$ is closed.*

PROOF: (1) follows from Theorem 2.1.21. (2) follows from (1) by taking complements. ☺

Let S be a metric space with metric ρ. The "distance between two sets" might be defined as:

$$\operatorname{dist}(A, B) = \inf \{\, \rho(a, b) : a \in A, b \in B \,\}.$$

This is not a metric, since for example $\operatorname{dist}(A, B) = 0$ can happen if $A \neq B$. This happens, for example, if $A \cap B \neq \varnothing$. Or even if A and B have an accumulation point in common. But it can happen in other ways too:

(2.2.18) EXERCISE. *Give an example of two disjoint nonempty closed sets with $\operatorname{dist}(A, B) = 0$.*

If one of the sets is compact, then we do get positive distance.

(2.2.19) THEOREM. *If A is closed, B is compact, and $A \cap B = \varnothing$, then $\operatorname{dist}(A, B) > 0$.*

PROOF: Suppose $\operatorname{dist}(A, B) = 0$. Then there exist points $x_n \in A$ and $y_n \in B$ with $\rho(x_n, y_n) < 1/n$. Now B is compact, so (by replacing the sequences with subsequences) we may assume that (y_n) converges. Say $y_n \to y \in B$. Then $x_n \to y$, also. But A is closed, so $y \in A$. Therefore $A \cap B \neq \varnothing$. ☺

For $A \subseteq S$ and $x \in S$, we may define $\operatorname{dist}(x, A) = \operatorname{dist}(\{x\}, A)$. Now $\{x\}$ is a compact set, so we have $x \in \overline{A}$ if and only if $\operatorname{dist}(x, A) = 0$.

Uniform continuity. Let $f\colon S \to T$ be a function. Let us recall a definition: f is **continuous** iff, for every $x \in S$ and every $\varepsilon > 0$, there exists $\delta > 0$ so that $f[B_\delta(x)] \subseteq B_\varepsilon(f(x))$. Compare that to this definition: f is **uniformly continuous** iff, for every $\varepsilon > 0$, there exists $\delta > 0$ so that for every $x \in S$, we have $f[B_\delta(x)] \subseteq B_\varepsilon(f(x))$. What is the difference: in the first case, δ is allowed to depend not only on ε, but also on x; but in the second case, while δ still depends on ε, it does not depend on x.

Clearly every uniformly continuous function is continuous. But the converse may fail:

(2.2.20) EXERCISE. *Let* $f: \mathbf{R} \to \mathbf{R}$ *be defined by* $f(x) = x^2$. *Then* f *is continuous, but not uniformly continuous.*

However, when S is compact, then the two kinds of continuity are the same:

(2.2.21) THEOREM. *Let* S *be a compact metric space, let* T *be a metric space, and let* $f: S \to T$ *be a function. If* f *is continuous, then* f *is uniformly continuous.*

PROOF: Let $\varepsilon > 0$ be given. Then by continuity, for every $x \in S$, there is a positive number $\delta(x)$ such that $f[B_{\delta(x)}(x)] \subseteq B_{\varepsilon/2}(f(x))$. Now the collection

$$\left\{ B_{\delta(x)/2}(x) : x \in S \right\}$$

is an open cover of S. By compactness, there is a finite subcover, say

$$B_{\delta(x_1)/2}(x_1), B_{\delta(x_2)/2}(x_2), \cdots, B_{\delta(x_n)/2}(x_n).$$

Let $\delta = \min\{\delta(x_1)/2, \delta(x_2)/2, \cdots, \delta(x_n)/2\}$. I claim that for this δ, and any $x \in S$, we have $f[B_\delta(x)] \subseteq B_\varepsilon(f(x))$. Indeed, suppose $y \in B_\delta(x)$. For some i, we have $y \in B_{\delta(x_i)/2}(x_i)$. Therefore $\rho(y, x_i) < \delta(x_i)/2$, so $\rho(f(y), f(x_i)) < \varepsilon/2$. But also $\rho(x, x_i) \le \rho(x, y) + \rho(y, x_i) < \delta(x_i)$, so $\rho(f(x), f(x_i)) < \varepsilon/2$. So we have $\rho(f(x), f(y)) < \varepsilon$. That is, we have $f[B_\delta(x)] \subseteq B_\varepsilon(f(x))$, so f is uniformly continuous. ☺

(2.2.22) THEOREM. *Let* S *be a compact metric space, and let* $\mathcal{U}$ *be an open cover of* S. *Then there is a positive number* r *such that for any set* $A \subseteq S$ *with* diam $A < r$, *there is a set* $U \in \mathcal{U}$ *with* $A \subseteq U$.

PROOF: First, there is a finite subcover, say $\{U_1, U_2, \cdots, U_n\}$. Suppose the assertion is false. Then for every $k \in \mathbf{N}$, there is a set A_k of diameter at most $1/k$ that is not contained in any U_i. So there are points $x_{ik} \in A_k \backslash U_i$, $(1 \le i \le n,$ $k \in \mathbf{N})$. By taking subsequences, we may assume that $\lim_k x_{ik} = x_i$ exists for each i. But the points x_i have distance 0 from each other, so they are all equal. This point x_i does not belong to any set U_i, contradicting the fact that the sets U_i cover S. ☺

The largest number r with the property of the Theorem is known as the **Lebesgue number** of the cover $\mathcal{U}$.

Number systems. Recall the situation from Section 1.6. Let b be a complex number, $|b| > 1$, and let D be a finite set of complex numbers, including 0. We are interested in representing complex numbers in the number system they define.

Write F for the set of "fractions"; that is numbers of the form

$$\sum_{j=1}^{\infty} a_j b^{-j}, \qquad a_j \in D.$$

(2.2.23) PROPOSITION. *The set F of fractions is a compact set.*

PROOF: Let $A \subseteq F$ be infinite. For each digit $d \in D$, write

$$A(d) = \left\{ x \in A : x = \sum_{j=1}^{\infty} a_j b^{-j}, a_1 = d \right\}$$

Then we have $A = \bigcup_{d \in D} A(d)$; since A is infinite, at least one of the sets $A(d)$ is also infinite. Choose $d_1 \in D$ so that $A(d_1)$ is infinite. Then write

$$A(d_1, d) = \left\{ x \in A : x = \sum_{j=1}^{\infty} a_j b^{-j}, a_1 = d_1, a_2 = d \right\}.$$

There is $d_2 \in D$ so that $A(d_1, d_2)$ is infinite. We may continue in this way: we obtain a sequence (d_j) of digits so that $A(d_1, d_2, \cdots, d_k)$ is infinite for all k. Then the number

$$\sum_{j=1}^{\infty} d_j b^{-j}$$

is an accumulation point of the set A. This shows that F is compact. ☺

2.3. UNIFORM CONVERGENCE

In Chapter 1 I had several occasions to mention "convergence" of a sequence of sets. There are two ways that will be used to make that idea precise. The first way, in this section, is "uniform convergence" of functions. The second way, in Section 2.4, is the metric of Hausdorff. In both cases, the most natural setting is that of an appropriate metric space. This is one of the reasons that we have been considering abstract metric spaces, rather than just subsets of Euclidean space.

Uniform convergence in general. Let S and T be two metric spaces. We will be considering functions $f : S \to T$.

Let f_n be a sequence of functions from S to T, and let f be another function from S to T. The sequence f_n **converges uniformly** (on S) to the function f iff for every $\varepsilon > 0$, there exists $N \in \mathbb{N}$ so that for all $x \in S$ and all $n \geq N$, we have $\rho(f_n(x), f(x)) < \varepsilon$.

This definition may be rephrased as convergence for a metric. We will say that two functions f and g are within **uniform distance** r of each other iff

$$\rho(f(x), g(x)) \leq r \qquad \text{for all } x \in S.$$

Then we write $\rho_u(f, g) \leq r$. More precisely, $\rho_u(f, g)$ is the smallest number r that works:

$$\rho_u(f, g) = \sup \{ \rho(f(x), g(x)) : x \in S \}.$$

Uniform convergence makes sense even if the functions are not continuous. Here is the most important property of uniform convergence:

(2.3.1) THEOREM. *Suppose* $f_n \colon S \to T$ *is a sequence of functions from the metric space* S *to the metric space* T. *Suppose* f_n *converges uniformly to a function* f. *If all functions* f_n *are continuous, then* f *is continuous.*

PROOF: Let $x \in S$ be given. I must show that f is continuous at x. So let a positive number ε be given. I must find a corresponding value δ. Now f_n converges to f uniformly, and $\varepsilon/3$ is a positive number, so there exists $N \in \mathbb{N}$ such that for all $n \geq N$, and all $y \in S$, we have $\rho\big(f(y), f_n(y)\big) < \varepsilon/3$. Once N is known, we can use the fact that the single function f_N is continuous at x to conclude that there exists $\delta > 0$ so that $\rho(y, x) < \delta$ implies $\rho\big(f_N(x), f_N(y)\big) < \varepsilon/3$. But then, for any y with $\rho(x, y) < \delta$, we have

$$\rho\big(f(x), f(y)\big) \leq \rho\big(f(x), f_N(x)\big) + \rho\big(f_N(x), f_N(y)\big) + \rho\big(f_N(y), f(y)\big)$$
$$< \frac{\varepsilon}{3} + \frac{\varepsilon}{3} + \frac{\varepsilon}{3} = \varepsilon.$$

This shows that f is continuous at x. ☺

If the set of functions we deal with is properly restricted, then ρ_u will be a metric. This is useful, since then all of the work we have done on metric spaces will be relevant. If S and T are metric spaces, we will write $\mathcal{C}(S, T)$ for the set of all continuous functions from S to T.

(2.3.2) THEOREM. *Let* S *be a compact metric space, and let* T *be a metric space. Then* ρ_u *is a metric on* $\mathcal{C}(S, T)$.

PROOF: First, $\rho_u(f, g) \geq 0$ and $\rho_u(f, g) = \rho_u(g, f)$ are clear. Since S is compact, we have $\rho_u(f, g) < \infty$, as in Corollary 2.2.16. If $f = g$, then $\rho_u(f, g) = 0$. If $\rho_u(f, g) = 0$, then $0 \leq \rho\big(f(x), g(x)\big) \leq 0$ for all $x \in S$, so $f(x) = g(x)$ for all $x \in S$, which means $f = g$. Finally, I must prove the triangle inequality. For any $x \in S$, we have

$$\rho\big(f(x), h(x)\big) \leq \rho\big(f(x), g(x)\big) + \rho\big(g(x), h(x)\big) \leq \rho_u(f, g) + \rho_u(g, h).$$

Therefore f and h are within uniform distance $\rho_u(f, g) + \rho_u(g, h)$ of each other, so $\rho_u(f, h) \leq \rho_u(f, g) + \rho_u(g, h)$. ☺

The metric will be called the **uniform metric**.

It makes sense to ask about the metric properties of the metric space $\mathcal{C}(S, T)$. Completeness will be a very useful property.

(2.3.3) THEOREM. *Suppose* S *is a compact metric space, and* T *is a complete metric space. Then the metric space* $\mathcal{C}(S, T)$ *is complete.*

PROOF: Let $(f_n)_{n \in \mathbb{N}}$ be a Cauchy sequence in $\mathcal{C}(S, T)$. Let $x \in S$. Then we have

$$\rho\big(f_n(x), f_m(x)\big) \leq \rho_u(f_n, f_m),$$

so $(f_n(x))_{n \in \mathbb{N}}$ is a Cauchy sequence in T. Since T is complete, this Cauchy sequence converges. Call its limit $f(x)$. This construction is valid for each $x \in S$, so it defines a function $f : S \to T$. I must show that f_n converges uniformly to f. Let $\varepsilon > 0$ be given. There is $N \in \mathbb{N}$ so that $\rho_u(f_n, f_m) < \varepsilon/2$ for all $n, m \geq N$. For any $x \in S$, there is $N_x \in \mathbb{N}$ so that $\rho\big(f_n(x), f(x)\big) \leq \varepsilon/2$ for $n \geq N_x$. We may assume $N_x \geq N$. Therefore, for any $n \geq N$, and any x, we have

$$\rho\big(f_n(x), f(x)\big) \leq \rho\big(f_n(x), f_{N_x}(x)\big) + \rho\big(f_{N_x}(x), f(x)\big) < \varepsilon.$$

Therefore $\rho_u(f_n, f) \leq \varepsilon$. This proves that f_n converges uniformly to f. By Theorem 2.3.1, the limit f is continuous, $f \in \mathcal{C}(S, T)$. So the space $\mathcal{C}(S, T)$ is complete under the uniform metric ρ_u. ☺

(2.3.4) EXERCISE. *Under what conditions on S and T is the metric space $\mathcal{C}(S, T)$ ultrametric?*

(2.3.5) EXERCISE. *Under what conditions on S and T is the metric space $\mathcal{C}(S, T)$ compact?*

Continuous curves. Suppose the metric space T is a convex subset of a Euclidean space $\mathbb{R}^k$, and S is a Euclidean space $\mathbb{R}^d$. A function $f : T \to S$ is **affine** iff it satisfies

$$f(tx + (1 - t)y) = tf(x) + (1 - t)f(y)$$

whenever $x, y \in T$ and $0 \leq t \leq 1$.

(2.3.6) PROPOSITION. *An affine map f of an interval $[u, v]$ into $\mathbb{R}$ must be of the form $f(x) = mx + b$, for some $m, b \in \mathbb{R}$.*

PROOF: If $u = v$, then of course $f(x) = f(u)$ for all $x \in [u, v]$. So suppose $u < v$. Now if $x \in [u, v]$, then

$$x = \frac{v - x}{v - u} u + \frac{x - u}{v - u} v,$$

which is of the form $tu + (1 - t)v$ with $0 \leq t \leq 1$. So

$$f(x) = \frac{v - x}{v - u} f(u) + \frac{x - u}{v - u} f(v)$$
$$= \frac{f(v) - f(u)}{v - u} x + \frac{vf(u) - uf(v)}{v - u},$$

which has the required form. ☺

(2.3.7) EXERCISE. *An affine map f of an interval $[u, v]$ into $\mathbb{R}^d$ must be of the form $f(x) = xm + b$, for some $m, b \in \mathbb{R}^d$.*

A **continuous curve** in a metric space S is a continuous function $f : [0, 1] \to S$. (Illustration in Figure 2.3.8.) Sometimes we will use the term to refer to the range $f\big[[0, 1]\big]$ of such a function.

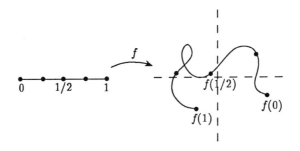

Figure 2.3.8. Continuous curve.

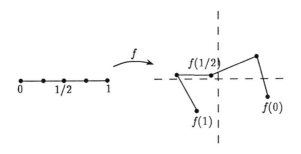

Figure 2.3.9. Piecewise affine curve.

A continuous curve $f: [0, 1] \to \mathbf{R}^d$ is **piecewise affine** iff there exist "division points" $0 = a_0 < a_1 < \cdots < a_n = 1$, such that f is affine on each of the subintervals $[a_{j-1}, a_j]$. The range of such a function is called a **polygonal curve**.

Consider the dragon construction of the Koch curve (page 18). There is a sequence (P_k) of polygonal curves. This sequence is supposed to "converge" to the Koch curve P. Since we do not have $P_{k+1} \subseteq P_k$, it is not reasonable to take the intersection. Here is one way to produce a reasonable limit.

The curve P_k consists of 4^k line segments. Divide the interval $[0, 1]$ into 4^k subintervals of equal length. Define a function $g_k: [0, 1] \to \mathbf{R}^2$, so that it is affine on each of these subintervals, and continuous on all of $[0, 1]$, so that the subintervals are mapped to the line segments of P_k. (To be explicit, start with $g_k(0)$ at the left end of P_k.) Thus we have $g_k[[0, 1]] = P_k$.

(2.3.10) PROPOSITION. *The sequence (g_k), in the dragon construction of the Koch curve, converges uniformly.*

PROOF: First note that every point of P_1 is within distance 1 of every point of P_0. So certainly $\rho_u(g_0, g_1) \le 1$. Now we estimate $\rho_u(g_k, g_{k+1})$. Let $t \in [0, 1]$. Then t is in one of the 4^k subintervals J on which g_k is affine. Its image $g_k[J]$ is a line segment of length 3^{-k}. When g_{k+1} is constructed, the interval J is subdivided into 4 equal intervals, and a miniature copy of g_1 is used there (properly translated, rotated, and dilated by 3^{-k}). So $\rho\big(g_k(t), g_{k+1}(t)\big) \leq 3^{-k}\rho(g_0, g_1)$. Therefore $\rho_u(g_k, g_{k+1}) \leq 3^{-k}$.

But then for $m > k$ we have

$$\rho_u(g_k, g_m) \leq \sum_{j=k}^{m-1} \rho_u(g_j, g_{j+1}) \leq \sum_{j=k}^{m-1} 3^{-j} < 3^{-k+1}/2.$$

This shows that (g_k) is a Cauchy sequence for the uniform metric, so it converges uniformly by Theorem 2.3.3. ☺

Write $g(t) = \lim_{k \to \infty} g_k(t)$ for all $t \in [0, 1]$. By Theorem 2.3.1, the function g is continuous. The limit set, the Koch curve, is then $P = g\big[[0, 1]\big]$. This also explains the use of the word "curve".

(2.3.11) EXERCISE. *Provide similar description for Heighway's dragon. Be sure to prove the uniform convergence that is required.*

Now that we have a rigorous definition for the "convergence" that is involved in the definitions, we may try to prove some of the properties of the dragon curves.

Is there a uniform convergence construction for the limit of the sequence (L_k) on page 5? Not in the same sense as for the dragons. The limit set (which is presumably the Cantor dust) *is not* a continuous curve. (We will prove this later, Proposition 3.1.4.) But there is a similar sort of uniform convergence construction. We must replace the interval $[0, 1]$ with a different parameter space.

Consider the space $E^{(\omega)}$ of infinite strings from the alphabet $\{\mathbf{0}, \mathbf{1}\}$. Define functions $g_k \colon E^{(\omega)} \to \mathbb{R}$ as follows. Begin with $g_0(\sigma) = 0$ for all σ. Then $g_0\big[E^{(\omega)}\big] = \{0\} = L_0$. Next, define $g_1(\mathbf{0}\sigma) = 0$ and $g_1(\mathbf{1}\sigma) = 2/3$. Then $g_1\big[E^{(\omega)}\big] = \{0, 2/3\} = L_1$. The function g_1 is continuous since the sets $[\mathbf{0}]$ and $[\mathbf{1}]$ are open sets. We may continue recursively: If g_k has been defined, then

$$g_{k+1}(\mathbf{0}\sigma) = \frac{g_k(\sigma)}{3}$$

$$g_{k+1}(\mathbf{1}\sigma) = \frac{g_k(\sigma) + 2}{3}.$$

The remainder is left to the reader:

(2.3.12) EXERCISE. *These functions $g_k \colon E^{(\omega)} \to \mathbb{R}$ are continuous and satisfy $g_k\big[E^{(\omega)}\big] = L_k$. The sequence (g_k) converges uniformly.*

(2.3.13) EXERCISE. *Show that the limit function $g = \lim g_k$ is the model map for the Cantor dust described on page 14. So $g\big[E^{(\omega)}\big]$ is exactly the triadic Cantor dust.*

Space-filling curves. A continuous curve $f \colon [0, 1] \to \mathbb{R}^d$, where $d \geq 2$ is called a **space-filling curve** iff $f\big[[0, 1]\big]$ contains a ball $B_r(x)$. This possibility was

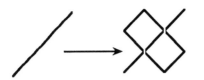

Figure 2.3.14. Construction for Peano curve.

realized by Peano, so it might be called a Peano curve. (We will see later that Heighway's dragon is a space-filling curve, Proposition 2.4.3.) Here is an easy example of a Peano curve. It can be done as a dragon curve. To go from one stage to the next, each line segment is replaced by nine segments with 1/3 the length, as in Figure 2.3.14. Figure 2.3.15 shows a few stages, as usual.

(2.3.15) EXERCISE. *Prove that this is a space-filling curve.*

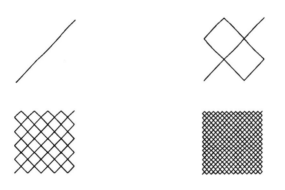

Figure 2.3.16. Peano's curve.

2.4. THE HAUSDORFF METRIC

Felix Hausdorff devised a way to describe convergence of sets. It is (for some purposes) better than the one discussed in the previous section, since it does not require finding an appropriate parameter space and parameterizations for the sets. It is simply a definition of a metric that applies to sets.

Convergence of sets. Let S be a metric space. Let A and B be subsets of S. We say that A and B are within **Hausdorff distance** r of each other iff every point of A is within distance r of some point of B, and every point of B is within distance r of some point of A.

This idea can be made into a metric, called the **Hausdorff metric**, D. If A is a set and $r > 0$, then the **open r-neighborhood** of A is

$$N_r(A) = \{\, y : \rho(x, y) < r \text{ for some } x \in A \,\}.$$

The definition of the Hausdorff metric D:

$$D(A, B) = \inf \{\, r : A \subseteq N_r(B) \text{ and } B \subseteq N_r(A) \,\}.$$

By convention, $\inf \varnothing = \infty$.

This definition D does not define a metric, however. There are various problems. For example, in $\mathbb{R}$, what is the distance between $\{0\}$ and $[0, \infty)$? It is infinite. That is not allowed in the definition of metric. Therefore, we will restrict the use of D to bounded sets. What is the distance $D(\varnothing, \{0\})$? Again, infinite. So we will restrict the use of D to nonempty sets. What is the distance $D((0, 1), [0, 1])$? Now the distance is 0, even though the two sets are not equal. Therefore we will restrict the use of D to closed sets. In fact for the purposes of this book, we will apply D only to nonempty compact sets.

If S is a metric space, we will write $\mathcal{K}(S)$ for the collection of all nonempty compact subsets of S.

(2.4.1) THEOREM. *Let S be a metric space. The Hausdorff function D is a metric on the set $\mathcal{K}(S)$.*

PROOF: First, clearly $D(A, B) \geq 0$ and $D(A, B) = D(B, A)$. Since S is compact, it is bounded, say $\operatorname{diam} S = r$. Then if $A \neq \varnothing$, we have $N_r(A) = S$. Therefore $D(A, B) < \infty$.

If $A = B$, then for every $\varepsilon > 0$ we have $A \subseteq N_\varepsilon(B)$; therefore $D(A, B) = 0$. Conversely, suppose $A, B \in \mathcal{K}(S)$ satisfy $D(A, B) = 0$. If $x \in A$, then for every $\varepsilon > 0$, we have $x \in N_\varepsilon(B)$, so $\operatorname{dist}(x, B) = 0$. Now B is compact, hence closed, so $x \in B$. This shows $A \subseteq B$. Similarly $B \subseteq A$, so $A = B$.

Finally we have the triangle inequality. Let $A, B, C \in \mathcal{K}(S)$. Let $\varepsilon > 0$. If $x \in A$, then there is $y \in B$ with $\rho(x, y) < D(A, B) + \varepsilon$. Then there is $z \in C$ with $\rho(y, z) < D(B, C) + \varepsilon$. This shows that A is contained in the $(D(A, B) + D(B, C) + 2\varepsilon)$-neighborhood of C. Similarly, C is contained in the $(D(A, B) + D(B, C) + 2\varepsilon)$-neighborhood of A. Therefore $D(A, C) \leq D(A, B) + D(B, C) + 2\varepsilon$. This is true for all $\varepsilon > 0$, so $D(A, C) \leq D(A, B) + D(B, C)$. ☺

Here is one way to describe the limit.

(2.4.2) EXERCISE. *Let A_n be a sequence of nonempty compact subsets of S. If A_n converges to A in the Hausdorff metric, then*

$$A = \{\, x : \text{there is a sequence } (x_n) \text{ with } x_n \in A_n \text{ and } x_n \to x \,\}.$$

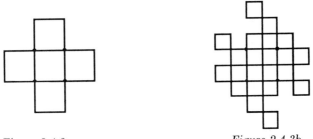

Figure 2.4.3a.   Figure 2.4.3b.

(2.4.3) PROPOSITION. *The Heighway dragon is a space-filling curve.*

PROOF: The information and notation used here are from the proof of Proposition 1.5.7. Suppose, at some stage of the construction, the polygon P_n contains all of the sides of a square S of the lattice L_n together with all of the sides of the four adjoining squares, as in Figure 2.4.3a. Then the polygon P_{n+2}, shown in Figure 2.4.3b, contains all four sides of all of the squares contained in S, together with all of the sides of all their adjoining squares.

Now the polygons P_7 and P_8 each contain such a square S. By induction, all sides of all of the subsquares of S are contained in the polygons P_k, for $k \geq 7$. By Exercise 2.4.2, the limit P contains all points of the interior of the square S. ☺

We may ask about the metric properties of the metric space $\mathcal{K}(S)$. The most important one will be completeness.

(2.4.4) THEOREM. *Suppose S is a complete metric space. Then the space $\mathcal{K}(S)$ is complete.*

PROOF: Suppose (A_n) is a Cauchy sequence in $\mathcal{K}(S)$. I must show that A_n converges. Let

$$A = \{\, x : \text{there is a sequence } (x_k) \text{ with } x_k \in A_k \text{ and } x_k \to x \,\}.$$

I will show that A_n converges to A.

Let $\varepsilon > 0$ be given. Then there is $N \in \mathbb{N}$ so that $n, m \geq N$ implies $D(A_n, A_m) < \varepsilon/2$. Let $n \geq N$. I claim that $D(A_n, A) \leq \varepsilon$.

If $x \in A$, then there is a sequence (x_k) with $x_k \in A_k$ and $x_k \to x$. So, for large enough k, we have $\rho(x_k, x) < \varepsilon/2$. Thus, if $k \geq N$, then (since $D(A_k, A_n) < \varepsilon/2$) there is $y \in A_n$ with $\rho(x_k, y) < \varepsilon/2$, and we have $\rho(y, x) \leq \rho(y, x_k) + \rho(x_k, x) \leq \varepsilon$. This shows that $A \subseteq N_\varepsilon(A_n)$.

Now suppose $y \in A_n$. Choose integers $k_1 < k_2 < \cdots$ so that $k_1 = n$ and $D(A_{k_j}, A_m) < 2^{-j}\varepsilon$ for all $m \geq k_j$. Then define a sequence (y_k) with $y_k \in A_k$

as follows: For $k < n$, choose $y_k \in A_k$ arbitrarily. Choose $y_n = y$. For $k_j < k <$ k_{j+1}, choose $y_k \in A_k$ with $\rho(y_{k_j}, y_k) < 2^{-j}\varepsilon$. Then y_k is a Cauchy sequence, so it converges. Let x be its limit. So $x \in A$. We have $\rho(y, x) = \lim_k \rho(y, y_k) < \varepsilon$. So $y \in N_\varepsilon(A)$. This shows that $A_n \subseteq N_\varepsilon(A)$.

So we have $D(A, A_n) \leq \varepsilon$. Therefore (A_n) converges to A. ☺

(2.4.5) EXERCISE. *Under what conditions on S is $\mathcal{K}(S)$ compact?*

(2.4.6) EXERCISE. *Under what conditions on S is $\mathcal{K}(S)$ ultrametric?*

Convergence in the examples. Whenever we have used the idea of "convergence" for a sequence of sets in a metric space, we have been talking about nonempty compact sets. The Hausdorff metric is the proper way to interpret it. In fact, it agrees with the other interpretations that have been used.

(2.4.7) PROPOSITION. *Let A_n be a sequence of nonempty compact sets, and suppose they decrease: $A_1 \supseteq A_2 \supseteq \cdots$. Then A_n converges to the intersection $A = \bigcap_{n \in \mathbb{N}} A_n$ in the Hausdorff metric.*

PROOF: Let $\varepsilon > 0$ be given. Now $A \subseteq A_n$, so $A \subseteq N_\varepsilon(A_n)$. For the other direction, note that the ε neighborhood of A,

$$N_\varepsilon(A) = \{\, y : \rho(x, y) < \varepsilon \text{ for some } x \in A \,\},$$

is an open set. The family

$$\{N_\varepsilon(A)\} \cup \{\, S \setminus A_n : n \in \mathbb{N} \,\}$$

is an open cover of A_1. Since A_1 is compact, there is a finite subcover. This means that, for some $N \in \mathbb{N}$, we have $(S \setminus A_n) \cup N_\varepsilon(A) \supseteq A_1$ for all $n \geq N$. Therefore we have $A_n \subseteq N_\varepsilon(A)$. So $D(A, A_n) \leq \varepsilon$ for all $n \geq N$. This shows that $A_n \to A$. ☺

(2.4.8) PROPOSITION. *Suppose S is a compact metric space, T is a metric space and the sequence $f_n : S \to T$ converges uniformly to f. Then the image sets $f_n[S]$ converge to $f[S]$ according to the Hausdorff metric on $\mathcal{K}(T)$.*

PROOF: This follows from the inequality $D(f[S], f_n[S]) \leq \rho_u(f, f_n)$. ☺

(2.4.9) EXERCISE. *Let A_n be a sequence of nonempty compact subsets of S. If A is a cluster point of the sequence (A_n), then A is contained in the set of all points $x \in S$ for which there exists a sequence (x_n) such that $x_n \in A_n$ and x is a cluster point of (x_n).*

2.5. Metrics for Strings

In Section 2.1, we defined a metric $\rho_{1/2}$ for the space $E^{(\omega)}$ of infinite strings from the two-letter alphabet $\{0, 1\}$. There are other, equally good, metrics for the same space.

Metrics for 01-strings. Let r be a real number satisfying $0 < r < 1$. A metric ρ_r is defined on $E^{(\omega)}$ in the same way as the metric $\rho_{1/2}$: if

$$\sigma = \alpha \sigma'$$
$$\tau = \alpha \tau',$$

where the first character of σ' is different than the first character of τ', and if $k = |\alpha|$ is the length of α, then

$$\rho_r(\sigma, \tau) = r^k.$$

(2.5.1) Exercise.

(1) ρ_r is a metric on $E^{(\omega)}$.
(2) The basic set $[\alpha]$ has diameter $r^{|\alpha|}$, for all $\alpha \in E^{(*)}$.
(3) The space $\left(E^{(\omega)}, \rho_r\right)$ is complete, compact, and separable.

(2.5.2) Proposition. *The metric spaces constructed from $E^{(\omega)}$ using the different metrics ρ_r are all homeomorphic to each other.*

Proof: Let $0 < r, s < 1$. If $h \colon E^{(\omega)} \to E^{(\omega)}$ is the identity function $h(\sigma) = \sigma$, then we will show that h is a homeomorphism from $\left(E^{(\omega)}, \rho_r\right)$ to $\left(E^{(\omega)}, \rho_s\right)$. It is enough to show that h is continuous, since applying this result with r and s interchanged will then show that h^{-1} is continuous.

Given $\varepsilon > 0$, choose k so that $s^k < \varepsilon$, then choose $\delta = r^k$. So if $\sigma, \tau \in E^{(\omega)}$ with $\rho_r(\sigma, \tau) < \delta$, then $\rho_r(\sigma, \tau) < r^k$, so σ and τ have at least their first k letters in common. But then $\rho_s(\sigma, \tau) \leq s^k < \varepsilon$. This shows that h is continuous. (In fact, uniformly continuous.) ☺

Since all of these metrics are homeomorphic, changing from one metric to another will make no difference for topological properties, such as compactness or separability, or such as the topological dimension (Chapter 3). So we will normally use the metric $\rho_{1/2}$ when dealing with topological properties. But changing from one metric to another does make a difference for metric properties, such as fractal dimension (Chapter 6). Here is a simple example where only one of the metrics ρ_r is appropriate: the model map for the Cantor dust (see page 14) has bounded distortion.

(2.5.3) PROPOSITION. *Let* $h \colon E^{(\omega)} \to \mathbf{R}$ *be the model map for the Cantor dust, satisfying*

$$h(\mathbf{0}\sigma) = \frac{h(\sigma)}{3}$$
$$h(\mathbf{1}\sigma) = \frac{h(\sigma) + 2}{3}.$$

Then we have

$$\frac{1}{3}\rho_{1/3}(\sigma, \tau) \le \big|h(\sigma) - h(\tau)\big| \le \rho_{1/3}(\sigma, \tau).$$

PROOF: Let σ, τ be given. Let α be their longest common prefix:

$$\sigma = \alpha\sigma',$$
$$\tau = \alpha\tau',$$

where the first letters of σ' and τ' are different. I will prove the inequalities by induction on the length $|\alpha|$. If $|\alpha| = 0$, that is α is the empty string Λ, then $\rho_{1/3}(\sigma, \tau) = 1$; the two real numbers $h(\sigma), h(\tau)$ both lie in $[0, 1]$, so $\big|h(\sigma) - h(\tau)\big| \le 1$; and σ, τ begin with different letters, so one of $h(\sigma), h(\tau)$ is in $[0, 1/3]$ and the other is in $[2/3, 1]$, and thus $\big|h(\sigma) - h(\tau)\big| \ge 1/3$. So the result is true when the length $|\alpha| = 0$.

Suppose the result is known when $|\alpha| = k$, and consider the case $|\alpha| = k + 1$. Then either $\alpha = \mathbf{0}\beta$ or $\alpha = \mathbf{1}\beta$ for some β with $|\beta| = k$. We will take the case $\alpha = \mathbf{1}\beta$; the other case is similar. By the induction hypothesis, applied to $\beta\sigma'$ and $\beta\tau'$,

$$\frac{1}{3}\rho_{1/3}(\beta\sigma', \beta\tau') \le \big|h(\beta\sigma') - h(\beta\tau')\big| \le \rho_{1/3}(\beta\sigma', \beta\tau').$$

But

$$\rho_{1/3}(\sigma, \tau) = \frac{1}{3}\,\rho_{1/3}(\beta\sigma', \beta\tau'),$$

and

$$\begin{aligned}
\big|h(\sigma) - h(\tau)\big| &= \big|h(\alpha\sigma') - h(\alpha\tau')\big| \\
&= \big|h(\mathbf{1}\beta\sigma') - h(\mathbf{1}\beta\tau')\big| \\
&= \left|\frac{h(\beta\sigma') + 2}{3} - \frac{h(\beta\tau') + 2}{3}\right| \\
&= \frac{1}{3}\big|h(\beta\sigma') - h(\beta\tau')\big|.
\end{aligned}$$

So the inequalities for this case follow. ☺

(2.5.4) EXERCISE. *If $r \neq 1/3$, then the model map h does not have bounded distortion from $\left(E^{(\omega)}, \rho_r\right)$ to $\mathbb{R}$.*

The range $h\left[\{0,1\}^{(\omega)}\right]$ of the function h in Proposition 2.5.3 is, of course, exactly the triadic Cantor dust C. The proposition shows that the space $\{0,1\}^{(\omega)}$ of infinite strings is homeomorphic to the Cantor dust C.

There are other (useful) ways to define a metric on the space $E^{(\omega)}$. Now let us assign a positive real number w_α to each node α of the binary tree $E^{(*)}$. If they satisfy the right conditions, then we will be able to define a metric ρ on $E^{(\omega)}$ so that the basic set $[\alpha]$ will have diameter exactly w_α.

(2.5.5) PROPOSITION. *Let a family w_α of positive real numbers be given, one for each node α in the infinite binary tree $E^{(*)}$. Suppose:*

$$w_\alpha > w_\beta \qquad \text{if } \alpha < \beta$$
$$\lim_{n \to \infty} w_{\sigma \restriction n} = 0 \qquad \text{for } \sigma \in E^{(\omega)}.$$

Then there is an ultrametric ρ on $E^{(\omega)}$ such that diam $[\alpha] = w_\alpha$ for all α.

PROOF: Define ρ as follows. If $\sigma = \tau$, then $\rho(\sigma, \tau) = 0$. If $\sigma \neq \tau$, then $\rho(\sigma, \tau) = w_\alpha$, where α is the longest common prefix of σ and τ.

Clearly $\rho(\sigma, \tau) \geq 0$. If $\sigma \neq \tau$, then $\rho(\sigma, \tau) = w_k > 0$. The equation $\rho(\sigma, \tau) = \rho(\tau, \sigma)$ is also easy.

So I must verify the ultra-triangle inequality,

$$\rho(\sigma, \tau) \leq \max\{\rho(\sigma, \theta), \rho(\theta, \tau)\}.$$

If two (or three) of the strings σ, θ, τ are equal, then this is trivial. So suppose they are all different. Let α be the longest common prefix of σ and θ; let β be the longest common prefix of θ and τ; let γ be the longest common prefix of σ and τ. Now α and β are both prefixes of θ, so one of them is a prefix of the other. If $\alpha \leq \beta$, then α is a prefix of both σ and τ, so $\alpha \leq \gamma$, and therefore

$$\rho(\sigma, \tau) = w_\gamma \leq w_\alpha = \rho(\sigma, \theta) \leq \max\{\rho(\sigma, \theta), \rho(\theta, \tau)\}.$$

The case $\beta \leq \alpha$ is similar.

Now I must show diam $[\alpha] = w_\alpha$. If $\sigma, \tau \in [\alpha]$, then the longest common prefix β of σ and τ is $\geq \alpha$, so $\rho(\sigma, \tau) = w_\beta \leq w_\alpha$. Therefore diam $[\alpha] \leq w_\alpha$. Choose any $\sigma \in E^{(\omega)}$; then $\alpha 0\sigma, \alpha 1\sigma \in [\alpha]$ and $\rho(\alpha 0\sigma, \alpha 1\sigma) = w_\alpha$. Therefore diam $[\alpha] \geq w_\alpha$. ☺

(2.5.6) EXERCISE. *Let ρ be the metric defined on $E^{(\omega)}$ using diameters w_α, where*

$$w_\alpha > w_\beta \qquad \text{if } \alpha < \beta$$

$$\lim_{n \to \infty} w_{\sigma \restriction n} = 0 \qquad \text{for } \sigma \in E^{(\omega)}.$$

Then

(1) $\left(E^{(\omega)}, \rho \right)$ *is homeomorphic to* $\left(E^{(\omega)}, \rho_{1/2} \right)$.
(2) *The countable set* $\left\{ [\alpha] : \alpha \in E^{(*)} \right\}$ *of "basic open sets" is equal to the set* $\left\{ B_r(\sigma) : \sigma \in E^{(\omega)}, r > 0 \right\}$ *of all open balls, and to the set of all closed balls.*

(2.5.7) PROPOSITION. *Let A be any subset of $E^{(\omega)}$ with at least two elements. Then there is a basic open set $[\alpha]$ such that $A \subseteq [\alpha]$ and* diam $A =$ diam $([\alpha])$.

PROOF: Let α be the longest common prefix for the set A (as in Proposition 1.3.2). The string α is finite (possibly empty) since A has at least two elements. Clearly $A \subseteq [\alpha]$, and therefore diam $A \leq$ diam $[\alpha]$. Now choose $\sigma \in A$. If k is the length $|\alpha|$, then of course $\sigma \restriction k = \alpha$. Since $\sigma \restriction (k+1)$ is not a common prefix for A, there is $\tau \in A$ with $\tau \restriction (k+1) \neq \sigma \restriction (k+1)$. That means that α is the longest common prefix for the pair $\{\sigma, \tau\}$, and therefore $\rho(\sigma, \tau) = w_\alpha$. Therefore diam $A \geq w_\alpha =$ diam $[\alpha]$. ☺

Other string spaces. It is probably clear to the reader by now that what has been done for strings from the alphabet $\{0, 1\}$ depends very little on that particular choice. Most of it could equally well be done for any (finite) alphabet.

Let E be any finite set; we assume that E has at least two elements. We call the elements of E "letters", and we call E an "alphabet".

(2.5.8) EXERCISE. *Formulate definitions of the following: string from the alphabet E; $E^{(n)}$; length $|\alpha|$ of a string α; $E^{(*)}$; $E^{(\omega)}$; $[\alpha]$. Formulate variants of the following results for this setting: 1.3.2, 2.1.8, 2.5.1, 2.5.2, 2.5.5, 2.5.6, 2.5.7.*

The "continuous" function is useful, but perhaps unfamiliar, on these string spaces. Roughly speaking, continuity of a function f means that any finite amount of information about $f(\sigma)$ is determined by only a finite amount of information about σ. Here is a more precise statement:

(2.5.9) EXERCISE. *Let E_1 and E_2 be two finite sets. Let $\sigma \in E_1^{(\omega)}$ and let $f \colon E_1^{(\omega)} \to E_2^{(\omega)}$ be a function. Then the following are equivalent:*

(1) f *is continuous at σ;*
(2) *for every integer n there exists m so that for $\tau \in E_1^{(\omega)}$, if we have $\tau \restriction m = \sigma \restriction m$ then we have $f(\tau) \restriction n = f(\sigma) \restriction n$.*

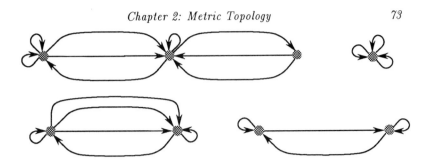

Figure 2.5.10. Directed multigraphs.

Path spaces of graphs. In Chapters 4 and 7 we will require a further generalization of the string spaces $E^{(\omega)}$.

Figure 2.5.10 shows examples of certain "graphs" that we will be using. There is a finite set V of **vertices** or **nodes**, and there are **edges**. Each edge goes from one vertex to another (or possibly back to the same one). The direction is important, so this is a **directed graph**. There may be more than one edge connecting a given pair of nodes, so this is a **multigraph**. This informal description (together with the pictures) is probably enough to tell you what a directed multigraph is. But we will need a more mathematically sound definition.

A **directed multigraph** consists of two (finite) sets V and E, and two functions $i \colon E \to V$ and $t \colon E \to V$. The elements of V are called **vertices** or **nodes**; the elements of E are called **edges** or **arrows**. For an edge e, we call $i(e)$ the **initial vertex** of e, and we call $t(e)$ the **terminal vertex** of e. We will often write E_{uv} for the set of all edges e with $i(e) = u$ and $t(e) = v$.

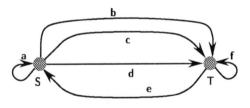

Figure 2.5.11. An example.

There is an example pictured in Figure 2.5.11. In this case, $V = \{\mathsf{S}, \mathsf{T}\}$ and $E = \{\mathsf{a}, \mathsf{b}, \mathsf{c}, \mathsf{d}, \mathsf{e}, \mathsf{f}\}$. We have, for example, $i(\mathsf{c}) = \mathsf{S}$, $t(\mathsf{c}) = \mathsf{T}$, $i(\mathsf{f}) = t(\mathsf{f}) = \mathsf{T}$.

A **path** in a directed multigraph is a sequence of edges, taken in some order, so that the terminal vertex of one edge is the initial vertex of the next edge. A path will often be identified with a string made up of the labels of the edges. Here are some examples of paths in the example: **c** or **bedfe** or **aaaa**. The

initial vertex of a path is, by definition, the initial vertex of the first edge in the path. Similarly, the **terminal vertex** of a path is the terminal vertex of the last edge in the path. We extend the functions i and t accordingly. For example, if $\alpha = $ **bedfe**, then $i(\alpha) = $ **S** and $t(\alpha) = $ **S**.

We will write $E_{uv}^{(*)}$ for the set of all paths with initial vertex u and terminal vertex v. We may say that such a path **goes from** u **to** v, or that it **connects** u to v. The number of edges in a path is its **length**, written $|\alpha|$. So in the example $\alpha = $ **bedfe**, we have $|\alpha| = 5$. We will write $E_{uv}^{(n)}$ for the set of all paths from u to v of length n; and $E_u^{(n)}$ for the set of all paths of length n with initial vertex u; and $E^{(n)}$ for the set of all paths of length n. The empty set conventions will work out best if we say (by convention) that for each $u \in V$, the set $E_{uu}^{(0)}$ has one element, the empty path from u to itself, written Λ_u. Of course, we may identify E with $E^{(1)}$ and E_{uv} with $E_{uv}^{(1)}$. Note that the conventions have been set so that if the strings α, β represent paths, and the terminal vertex of α is equal to the initial vertex of β, then the concatenated string $\alpha\beta$ represents a path, as well.

The case that we have considered before, where all strings from a given alphabet E are allowed, corresponds to a graph with only one vertex, and the alphabet E as edge set.

A path α with $i(\alpha) = t(\alpha)$ is called a **cycle**. A cycle of length 1 is a **loop**. A directed multigraph is **strongly connected** iff, for each pair u, v of vertices, there is a path from u to v.

Let (V, E, i, t) be a directed multigraph. We will consider the set $E^{(*)}$ of all paths in the graph. This naturally has the structure of a "tree": If α is a path, then the children of α are the paths αe, for edges e with $i(e) = t(\alpha)$. Actually this is not a tree: it is a finite disjoint union of trees, one tree $E_v^{(*)}$ corresponding to each node v of the graph. A disjoint union of trees is sometimes called a **forest**. So we will call this the **path forest** of the graph.

What about infinite paths, corresponding to the infinite strings that we have considered before? An infinite string σ corresponds to an infinite path if the terminal vertex for each edge matches the initial vertex for the next edge. We write $E^{(\omega)}$ for the set of all infinite paths for the graph (V, E, i, t). If $v \in V$ is a vertex, then we write $E_v^{(\omega)}$ for the set of all infinite paths starting at v. (It usually does not make sense to assign a terminal vertex to an infinite path.) There will be one of these path spaces $E_v^{(\omega)}$ for each vertex of the graph (V, E, i, t), or one for each tree $E_v^{(*)}$ in the forest $E^{(*)}$ of finite paths. If $\alpha \in E^{(*)}$, then we write $[\alpha] = \{ \sigma \in E^{(\omega)} : \alpha \le \sigma \}$, that is, the set of all infinite paths that begin with the finite path α.

Metrics may be defined on the path spaces $E_v^{(\omega)}$ in much the same way as has already been done. Some notice must be taken of the possibility that some nodes in $E^{(*)}$ have no children, or only one child. Now if α has no children, then $[\alpha] = \varnothing$, so its diameter must be 0. If α has only one child β, then $[\alpha] = [\beta]$, so diam $[\alpha]$ must be equal to diam $[\beta]$. These cases will not occur if, in the graph (V, E, i, t), each node has at least two edges leaving it.

Let (V, E, i, t) be a directed multigraph. Let a family w_α of positive real numbers be given, one for each node α in the path forest $E^{(*)}$. We want to define distances $\rho(\sigma, \tau)$. In fact, only distances between infinite strings with the same initial vertex will be needed. So we can think of defining several (disjoint) metric spaces $E_v^{(\omega)}$, one for each node $v \in V$. We will normally write simply ρ for all the metrics involved.

Suppose the family w_α satisfies

$$w_\alpha > w_\beta \qquad \text{if } \alpha < \beta$$
$$\lim_{n \to \infty} w_{\sigma \restriction n} = 0 \qquad \text{for } \sigma \in E^{(\omega)}.$$

The definition for the metrics is as before. If $\sigma, \tau \in E_v^{(\omega)}$, and $\sigma \neq \tau$, then the two strings have a longest common prefix, since they have at least the prefix Λ_v in common. So define $\rho(\sigma, \tau) = w_\alpha$, where α is the longest common prefix of σ and τ. This defines ultrametrics ρ on the spaces $E_v^{(\omega)}$ such that $\operatorname{diam}[\alpha] = w_\alpha$ for "most" α. For which α might the equation fail?

(2.5.12) EXERCISE. *Adapt the following results to this setting: 2.5.6, 2.5.7.*

*2.6. REMARKS

It is probably more conventional in fractal geometry to limit consideration to subsets of Euclidean space. But metric spaces are the proper setting for the theory of topological dimension, and perhaps even for the theory of Hausdorff dimension. We will need uniform convergence of functions, and the Hausdorff metric for sets; even if the primary interest is Euclidean space, these lead naturally to other metric spaces.

On the other hand, I have limited the discussion to metric spaces, rather than more general topological spaces. Certainly Hausdorff dimension belongs in a metric space. Topological dimension can be done in much greater generality, but I will leave that to the student who wants to go beyond this book. In fact, whenever it is convenient, the discussion will be restricted to separable metric spaces, or even to compact metric spaces.

Additional material along the lines of this chapter can be found in texts called "general topology" or "point-set topology", such as [32] or [35].

According to the definition in this chapter, d-dimensional Euclidean space is the set $\mathbf{R}^d$ of d-tuples of real numbers. For our purposes, this is reasonable. But in some other contexts, one would say that $\mathbf{R}^d$ is really more properly considered to be d-dimensional Euclidean space *together with* a distinguished coordinate system. This sort of distinction emphasizes the difference between two camps called "synthetic geometry" and "analytic geometry".

---

*An optional section.

The reader aware of such things may have noticed that the axiom of choice is freely used in this chapter. There is no better way. If we want to do metric topology using sequences, then at least countable choice must be used.

The spaces $E^{(\omega)}$ that have been called here "string models" or "path models" are more commonly known as "shift models" and "subshift models". However, the word "shift" involved refers to the left shift on these spaces, which is never† mentioned in this book, so its use seemed inappropriate.

For the Hausdorff metric on sets in a metric space see for example [28, §28]. Felix Hausdorff's mathematical writings are few in number, but immensely influential.

Comments on exercises. Do not read this until you have tried to do the exercise yourself!

Exercise 2.1.2: $x \leq y \leq z$ or $x \geq y \geq z$. That is, y is between x and z.

Exercise 2.1.7: $|x + y| = |x| + |y|$ if and only if one of x, y is a nonnegative scalar multiple of the other.

Exercise 2.1.14: An "if and only if" proof is done in two parts, an "if" part and an "only if" part. Two sets are equal when each is a subset of the other.

Exercise 2.1.25: This can be done from the ε, δ definition (good practice using the definition). Or it can be done using Theorem 2.1.24.

Exercise 2.1.30: Given a Cauchy sequence in $\mathbb{R}^3$, use the completeness of $\mathbb{R}$ three times; successively extract subsequences so that each of the three coordinates converges. Then show that the resulting subsequence converges.

Exercise 2.1.33: The sets $\mathbb{R}$ and $(0, 1)$ are homeomorphic.

Exercise 2.2.12: Show that $E^{(\omega)}$ is countably compact. If A is infinite, then at least one of the sets $A \cap [\mathbf{0}]$, $A \cap [\mathbf{1}]$ is infinite. Then at least one of the subsets $A \cap [\alpha]$, $\alpha \in E^{(2)}$, is infinite. Etc.

Exercise 2.2.18: In the plane, consider the two graphs $xy = 1$ and $xy = -1$.

Exercise 2.4.2: Suppose $A_n \to A$, and let

$$B = \{\, x : \text{there is a sequence } (x_n) \text{ with } x_n \in A_n \text{ and } x_n \to x \,\}.$$

First, $A \subseteq B$: Since any element $x \in A$ is within distance $D(A, A_n) + 1/n$ of some element x_n of A_n, we have $x_n \to x$. For $B \subseteq A$: Since any element of B is within distance ε of elements of A_n for large enough n, and those elements, in turn, are within distance ε of elements of A. Then apply 2.2.18.

> I have, with difficulty, been prevented
> by my friends from labeling [the book]:
> What Every Young Analyst Should Know.
> —J. L. Kelley, *General Topology*

---

†With this exception

3

Topological Dimension

The sets of elementary geometry have associated with them a **dimension**. Points have dimension 0. Curves have dimension 1. Surfaces have dimension 2. Solids have dimension 3. When we leave elementary geometry, there is the possibility of considering point sets not falling into any of these clear-cut groups. Mathematicians have defined a general notion of dimension to help out in the study of such sets. In fact, they have done it in several different ways; one way of defining dimension may be useful for one purpose but not for another purpose.

We will discuss more than one of these definitions. The definitions that will be considered fall generally into two broad classes: topological dimension and fractal dimension. The first one is a topological dimension known as the "small inductive dimension".

3.1. SMALL INDUCTIVE DIMENSION

Hermann Weyl explained dimension like this:* "We say that space is 3-dimensional because the walls of a prison are 2-dimensional."

If we have a point that we want to imprison, we can use a small cube as the prison. By making the cube small enough, when the point is forbidden to move through the faces of the cube, it can be confined to a very small region. The cube consists of 6 plane faces; we need to know that they are 2-dimensional. A point living in these faces (Flatland) can be imprisoned using a small circle. So saying that the faces of the cube are 2-dimensional requires knowing that a circle is 1-dimensional. A point living in the circle (Lineland) can be imprisoned using just 2 points as prison walls. So we need to know that a 2-point set is 0-dimensional. Finally, a point living in the 2-point set (Pointland) is already unable to move. So we need no prison walls at all. This will be the definition of a 0-dimensional set.

---

*I have to quote from memory, because I cannot seem to find this quotation now.

The definition. A subset of a metric space is **clopen** iff it is both closed and open. A metric space is called **zero-dimensional** iff there is a base for the open sets consisting of clopen sets. Zero-dimensionality is a topological property: If the metric spaces S and T are homeomorphic, and one of them is zero-dimensional, then so is the other.

A good example of a zero-dimensional metric space is an ultrametric space. In an ultrametric space, every open ball is clopen, and the collection of all open balls is a base for the open sets.

Another easy example is a finite set. If $S = \{x_1, x_2, \cdots, x_n\}$, and we let ε be the minimum of the numbers $\rho(x_i, x_j)$ for $i \neq j$, then $\varepsilon > 0$. Each of the balls $B_\varepsilon(x_j)$ is clopen, and these balls constitute a base for the open sets of S.

The next example to consider is the line $\mathbb{R}$. It is not zero-dimensional. To prove this, I will need to show that there are not very many clopen sets in $\mathbb{R}$. Note that is is *not* enough merely to show that the elements $B_r(x)$ of the usual base are not clopen: it is conceivable that even if these are not clopen, there may be some *other* base for the open sets that is made up of clopen sets.

(3.1.1) THEOREM. *The only clopen sets in the space $\mathbb{R}$ are $\varnothing$ and $\mathbb{R}$. Therefore, $\mathbb{R}$ is not zero-dimensional.*

PROOF: Let $A \subseteq \mathbb{R}$, and suppose $A \neq \varnothing$ and $A \neq \mathbb{R}$. I must show that A is not a clopen set, or, equivalently, that A has a boundary point. We will define recursively two sequences, (x_n) and (y_n). First, we may choose a point $x_0 \in A$ since $A \neq \varnothing$. Also, we may choose a point $y_0 \notin A$ since $A \neq \mathbb{R}$. After x_n and y_n have been defined, with $x_n \in A$ and $y_n \notin A$, we want to define x_{n+1} and y_{n+1}. Consider the midpoint $z_n = (x_n + y_n)/2$. If $z_n \in A$, then define $x_{n+1} = z_n$, $y_{n+1} = y_n$; and if $z_n \notin A$, then define $x_{n+1} = x_n$, $y_{n+1} = z_n$. So in any case, we get $x_{n+1} \in A$ and $y_{n+1} \notin A$, with $|x_{n+1} - y_{n+1}| = |x_n - y_n|/2$. So by induction $|x_n - y_n| = |x_0 - y_0|/2^n$. Thus $|x_n - y_n| \to 0$ as $n \to \infty$. Also, $|x_{n+1} - x_n| \leq |x_n - y_n| = |x_0 - y_0|/2^n$, so (x_n) is a Cauchy sequence. Let $x = \lim_n x_n$. Because $|x_n - y_n| \to 0$, we have also $y_n \to x$. Therefore x is a boundary point of A. So A is not a clopen set. ☺

It is worth pointing out the following, which has the same proof.

(3.1.2) COROLLARY. *Let $a < b$. The only clopen subsets of the space $[a, b]$ are $\varnothing$ and $[a, b]$. Therefore, $[a, b]$ is not zero-dimensional.*

Next we show that the triadic Cantor dust is zero-dimensional. Recall that the Cantor dust C was defined as an intersection $C = \bigcap_{k \in \mathbb{N}} C_k$, and each C_k consists of 2^k disjoint closed intervals of length 3^{-k}. Write I_{kj} for those intervals, so that

$$C_k = \bigcup_{j=1}^{2^k} I_{kj}.$$

(3.1.3) PROPOSITION. *The sets $M_{kj} = C \cap I_{kj}$ $(k = 0, 1, \cdots; j = 1, 2, \cdots, 2^k)$ constitute a base for the open sets of C. They are clopen in C, so C is zero-dimensional.*

PROOF: The interval I_{kj} has length 3^{-k}, and the distance from I_{kj} to any other interval $I_{kj'}$ is at least 3^{-k}. So if $I_{kj} = [a, b]$, then we see that $M_{kj} = C \cap [a, b]$ is closed in C, since $[a, b]$ is closed in $\mathbb{R}$; and $M_{kj} = C \cap (a - 3^{-k}, b + 3^{-k})$ is open in C since $(a - 3^{-k}, b + 3^{-k})$ is open in $\mathbb{R}$.

If $x \in C$ and $\varepsilon > 0$, then choose k so that $3^{-k} < \varepsilon$, and j so that $x \in I_{kj}$. Then $C \cap B_\varepsilon(x) \supseteq M_{kj}$. This shows that the collection of all M_{kj} is a base for the open sets. ☺

Here is a result promised in Chapter 2:

(3.1.4) PROPOSITION. *Let C be the Cantor dust. There is no continuous function from the interval $[0, 1]$ onto C.*

PROOF: Suppose $h: [0, 1] \to C$ is surjective. The set M_{11} is a clopen subset of C. Both M_{11} and its complement $C \setminus M_{11} = M_{12}$ are nonempty. So both the set $h^{-1}[M_{11}]$ and its complement $h^{-1}[M_{12}]$ are nonempty. By Corollary 3.1.2, not both sets are open. Therefore, by Theorem 2.1.21, h is not continuous. ☺

A metric space S has **small inductive dimension 1** iff S is not zero-dimensional and there is a base for the open sets consisting of sets with zero-dimensional boundary.

(3.1.5) THEOREM. *The line $\mathbb{R}$ has small inductive dimension 1.*

PROOF: The usual base for the line consists of balls $B_r(x) = (x - r, x + r)$. The boundary of such a ball is a two-point set $\{x - r, x + r\}$, which is zero-dimensional. By Theorem 3.1.1, the line is not zero-dimensional. Therefore $\mathbb{R}$ has small inductive dimension 1. ☺

In general, the **small inductive dimension** for metric spaces is defined in an inductive manner. Each metric space S will be assigned a dimension, written ind S, chosen from the set $\{-1, 0, 1, 2, 3, \cdots, \infty\}$, consisting of the integers ≥ -1 together with an extra symbol ∞, considered to be larger than all of the integers. The empty metric space $\varnothing$ has ind $\varnothing = -1$. If k is a nonnegative integer, then we say that ind $S \leq k$ iff there is a base for the open sets of S consisting of sets U with ind $\partial U \leq k - 1$.† We say ind $S = k$ iff ind $S \leq k$ but ind $S \not\leq k - 1$. Finally, if ind $S \leq k$ is false for all integers k, then we say ind $S = \infty$.

Since the small inductive dimension is defined inductively, it is often possible to prove things about it by induction. We will carry out the first few such results in detail.

---

†Recall the notation ∂U for the boundary of U.

(3.1.6) THEOREM. *Topological dimension is a topological property: If S and T are homeomorphic, then* ind $S =$ ind T.

PROOF: The proof is by induction on ind S. If ind $S = -1$, then S is empty; since T is homeomorphic to S, it is also empty, so ind $T = -1$.

Suppose the theorem is known for spaces S with ind $S \leq k$, and consider a space S with ind $S = k + 1$. Let $h \colon S \to T$ be a homeomorphism. There is a base $\mathcal{B}$ for the open sets of S consisting of sets B with ind $\partial B \leq k$. Now $\{\, h[B] : B \in \mathcal{B} \,\}$ is a base for the open sets of T. If $B \in \mathcal{B}$, then $h[\partial B] = \partial h[B]$. The restriction of h to ∂B is a homeomorphism. By the induction hypothesis, ind $\partial h[B] =$ ind $\partial B \leq k$. So we see that there is a base for the open sets of T consisting of sets with boundary of dimension $\leq k$. This shows that ind $T \leq k + 1$. But if ind $T \leq k$, then the induction hypothesis would show ind $S \leq k$, which is false. So ind $T = k + 1$. Therefore, by induction we see that if ind S is an integer, then ind $S =$ ind T.

If ind $S = \infty$, then ind $T = k$ is false for all integers k, so also ind $T = \infty$. So in all cases we have ind $S =$ ind T. ☺

(3.1.7) THEOREM. *Let S be a metric space, and let $T \subseteq S$. Then* ind $T \leq$ ind S.

PROOF: This is clear if ind $S = \infty$. So suppose ind $S < \infty$. The proof proceeds by induction on ind S.

If ind $S = -1$, then S is empty, so clearly $T \subseteq S$ is also empty, and hence ind $T = -1$.

Suppose the theorem is true for all pairs S, T with $T \subseteq S$ and ind $S \leq k$. Consider a pair $T \subseteq S$ with ind $S = k + 1$. I must show that there is a base for the open sets of T consisting of sets with boundary of dimension $\leq k$. So let $x \in T$, and let V be an open set in T with $x \in V$. I must find an open set U in T with $x \in U \subseteq V$ and ind $\partial_T U \leq k$. [Note that the boundary of a set in T may be different than the boundary of the same set in S; so the space is indicated as a subscript.] Now since V is open in T, there exists a set $\widetilde{V}$ open in S with $V = \widetilde{V} \cap T$. Since ind $S \leq k + 1$, and $x \in \widetilde{V}$, there is a set $\widetilde{U}$ open in S with $x \in \widetilde{U} \subseteq \widetilde{V}$ and ind $\partial_S \widetilde{U} \leq k$. Let $U = \widetilde{U} \cap T$. Then U is open in T, and $x \in U \subseteq V$. Now by Theorem 2.1.35, $\partial_T U \subseteq \partial_S \widetilde{U}$, so by the induction hypothesis, we have ind $\partial_T U \leq$ ind $\partial_S \widetilde{U} \leq k$. Thus there is a base for the open sets of T consisting of sets U with ind $\partial_T U \leq k$. This means that ind $T \leq k + 1$.

Therefore, by induction, the theorem is true for all values of ind S. ☺

(3.1.8) EXERCISE. *The Sierpiński gasket has small inductive dimension 1.*

The triadic Cantor dust C is homeomorphic to the space $\{0, 1\}^{(\omega)}$ of infinite strings based on the two-letter alphabet $\{0, 1\}$ (Proposition 2.5.3). Any metric space homeomorphic to these spaces may be called generically a "Cantor dust" or a "Cantor set". The Cantor dust is a **universal zero-dimensional space** in the following sense:

(3.1.9) THEOREM. *Let S be a nonempty separable metric space. Then* ind $S = 0$ *if and only if S is homeomorphic to a subset of the space $\{0,1\}^{(\omega)}$.*

PROOF: Suppose S is homeomorphic to $T \subseteq \{0,1\}^{(\omega)}$. By Theorem 3.1.6, ind $S = $ ind T. By Theorem 3.1.7, ind $T \leq$ ind $\{0,1\}^{(\omega)}$. But ind $\{0,1\}^{(\omega)} = 0$. Therefore ind $S \leq 0$. Since $S \neq \varnothing$, we have ind $S = 0$.

Conversely, suppose ind $S = 0$. There is a base $\mathcal{B}_1$ for the open sets of S consisting of clopen sets. By Theorem 2.2.1, there is a countable $\mathcal{B} \subseteq \mathcal{B}_1$, still a base for the open sets. Write $\mathcal{B} = \{U_1, U_2, \cdots\}$. For notation, we will use $U_i(1) = U_i$ and $U_i(0) = S \setminus U_i$; they are all clopen sets. If $\alpha \in \{0,1\}^{(k)}$, say $\alpha = e_1 e_2 \cdots e_k$, let $U(\alpha) = U_1(e_1) \cap U_2(e_2) \cap \cdots \cap U_k(e_k)$.

Define a map $h: S \to \{0,1\}^{(\omega)}$ as follows: given $x \in S$, the ith letter of $h(x)$ is 0 or 1 according as x belongs to $U_i(0)$ or $U_i(1)$. So this means $h(x) \in [\alpha]$ if and only if $x \in U(\alpha)$. Or $h^{-1}[[\alpha]] = U(\alpha)$.

Now I claim h is one-to-one. Indeed, if $x \neq y$, then $S \setminus \{y\}$ is an open set containing x, so there is i with $x \in U_i \subseteq S \setminus \{y\}$, and therefore $h(x) \neq h(y)$. This shows that h is one-to-one. So the inverse function $h^{-1}: h[S] \to S$ exists. Now the sets $[\alpha]$ constitute a base for the open sets of $\{0,1\}^{(\omega)}$, and $h^{-1}[[\alpha]] = U(\alpha)$ is open for every α, so h is continuous. Similarly, the sets U_i constitute a base for the open sets of S, and

$$h[U_i] = \bigcup_{\alpha \in \{0,1\}^{(i-1)}} \left(h[S] \cap [\alpha 1] \right)$$

is open in $h[S]$ for every i. So h^{-1} is continuous. This means that h is a homeomorphism of S onto $h[S] \subseteq \{0,1\}^{(\omega)}$. ☺

Number systems. Recall the situation from Section 1.6. Let b be a real number, $|b| > 1$, and let D be a finite set of real numbers, including 0. We are interested in representing real numbers in the number system they define.

Write W for the set of "whole numbers"; that is, numbers of the form

(1) $$\sum_{j=0}^{M} a_j b^j.$$

Write F for the set of "fractions"; that is numbers of the form

(2) $$\sum_{j=-\infty}^{-1} a_j b^j.$$

The set of all numbers represented by this system is the sum of one of each:

(3) $$\sum_{j=-\infty}^{M} a_j b^j.$$

Exercise 1.6.3 may be considered a problem on topological dimension.

(3.1.10) PROPOSITION. *Let b be a real number with $|b| > 1$, and let D be a finite set of real numbers, including 0. Then either some real number has no expansion in the form (3) or some real number has more than one expansion in the form (3).*

PROOF: Suppose every real number has a unique expansion in the form (3). In fact, it would follow that $\mathbb{R}$ is zero-dimensional. I will prove only that $\mathbb{R}$ has a clopen subset other than $\varnothing$ and $\mathbb{R}$, which contradicts Theorem 3.1.1.

First I claim that there is a minimum distance between elements of W. Suppose not: then choose $x_n, y_n \in W$ with $x_n \neq y_n$ and $|x_n - y_n| \to 0$. By subtracting the places where they agree, we may assume that x_n and y_n have only zeros in places higher than the highest power of b where they disagree. Let M_n be the highest power of b where the expansions of x_n and y_n disagree. Then $x'_n = x_n b^{-M_n - 1}$ and $y'_n = y_n b^{-M_n - 1}$ are elements of F, they differ in the first place to the right of the radix point, and still $|x'_n - y'_n| \to 0$. By taking subsequences, we may assume that the first place of x'_n is the same for all n, the first place of y'_n is the same for all n, (x'_n) converges, and (y'_n) converges. The limits $x = \lim x'_n$ and $y = \lim y'_n$ are equal, yet have different first places. This contradicts the uniqueness. So there is a minimum distance between elements of W.

Now I claim that the set F is clopen. If not, there is a boundary point x. Now F is closed, so $x \in F$. There is a sequence in $\mathbb{R} \setminus F$ that converges to x. So there exist $y_n \in F$ and $w_n \in W$ with $w_n \neq 0$ and $y_n + w_n \to x$. By compactness of F we may assume that (y_n) converges, say to $y \in F$. Then $w_n \to x - y$. Since there is a minimum distance between elements of W, this means that $w_n = x - y$ for large enough n. So we have $x = y + w_n$, contradicting uniqueness of representations.

Thus F is clopen. But $0 \in F$ so $F \neq \varnothing$; and F is compact so $F \neq \mathbb{R}$. This contradicts Theorem 3.1.1. ☺

3.2. LARGE INDUCTIVE DIMENSION

Separation of sets. An important consideration for topological dimension is "separation" of sets. The first theorem of this type holds in any metric space.

(3.2.1) THEOREM. *Let A and B be disjoint closed subsets of a metric space S. Then there exist disjoint open sets U and V in S with $U \supseteq A$ and $V \supseteq B$.*

PROOF: Write $U = \{ x \in S : \text{dist}(x, A) < \text{dist}(x, B) \}$. By the triangle inequality, we have

$$| \text{dist}(x, A) - \text{dist}(y, A)| \leq \rho(x, y),$$

so that $\text{dist}(x, A)$ is a continuous function of x. Similarly, $\text{dist}(x, B)$ is a continuous function of x. It follows that U is an open set. Since A is closed, we have $\text{dist}(x, A) = 0$ if and only if $x \in A$. So if $x \in A$, we have $\text{dist}(x, A) = 0 < \text{dist}(x, B)$; this shows that $A \subseteq U$. Let $V = \{ x \in S : \text{dist}(x, A) > \text{dist}(x, B) \}$. As before, V is open and $B \subseteq V$. Clearly $U \cap V = \varnothing$. ☺

This result can be rephrased in another form:

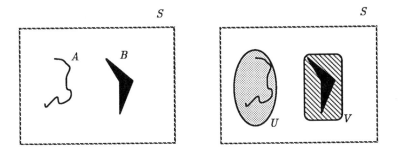

Figure 3.2.1. Illustration for Theorem 3.2.1.

(3.2.2) COROLLARY. *Suppose F is closed and U is open. If $F \subseteq U$, then there is an open set V with $F \subseteq V \subseteq \overline{V} \subseteq U$.*

The same idea of proof can be used to prove some variants of the theorem:

(3.2.3) EXERCISE.

(1) Let A and B be subsets of a metric space S. Suppose $\overline{A} \cap B = \varnothing = A \cap \overline{B}$. Then there exist disjoint open sets U and V in S with $U \supseteq A$ and $V \supseteq B$.

(2) Let A and B be disjoint closed subsets of a metric space S. Prove that there exist open sets U and V in S with $U \supseteq A$, $V \supseteq B$, and $\overline{U} \cap \overline{V} = \varnothing$. Rephrase this in the manner of Corollary 3.2.2.

If we know about the topological dimension, then a stronger conclusion can be obtained. (Notice the hypothesis of separability, however.)

(3.2.4) THEOREM. *Let A and B be disjoint closed subsets of a separable metric space S. Suppose ind $S = 0$. Then there is a clopen set U in S with $U \supseteq A$ and $U \cap B = \varnothing$.*

PROOF: Since ind $S = 0$, there is a base $\mathcal{B}$ for the open sets consisting of clopen sets. For each $x \in S$, choose a set $U(x) \in \mathcal{B}$ with $x \in U(x)$, and either $U(x) \cap A = \varnothing$ or $U(x) \cap B = \varnothing$. This is possible, since if $x \notin A$, then $S \setminus A$ is an open set containing x; if $x \notin B$, then $S \setminus B$ is an open set containing x; and every $x \in S$ satisfies at least one of these. Thus $\{ U(x) : x \in S \}$ is an open cover of S. But S is separable, so, by Theorem 2.2.1, there is a countable subcover, say

$$\{ U(x_1), U(x_2), \cdots \}.$$

Define $V_1 = U(x_1)$, and recursively $V_n = U(x_n) \setminus (V_1 \cup V_2 \cup \cdots \cup V_{n-1})$. These sets V_n are again clopen, and

$$\bigcup_{n \in \mathbb{N}} V_n = \bigcup_{n \in \mathbb{N}} U(x_n) = S.$$

Also, it is still true that each of the sets V_n is either disjoint from A or disjoint from B. Now write U for the union of all the sets V_n that are disjoint from B:

$$U = \bigcup \{ V_n : V_n \cap B = \varnothing \}.$$

Thus U is open and $U \cap B = \varnothing$. But the complement of U is

$$V = S \setminus U = \bigcup \{ V_n : V_n \cap B \neq \varnothing \}.$$

Thus V is open, so U is closed. Also, $V \cap A = \varnothing$, so $U \supseteq A$. ☺

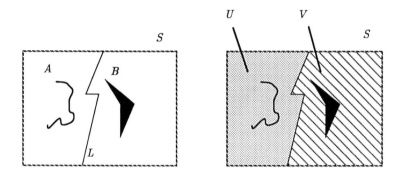

Figure 3.2.5. L separates A and B.

Let A and B be disjoint subsets of a metric space S. We say that a set $L \subseteq S$ **separates** A and B iff there exist open sets U and V in S with $U \cap V = \varnothing$, $U \supseteq A$, $V \supseteq B$, and $L = S \setminus (U \cup V)$. (See Figure 3.2.5.) So Theorem 3.2.4 says that in a zero-dimensional space S, any two disjoint closed sets are separated by the empty set.

The **large inductive dimension** is a topological dimension closely related to the small inductive dimension. Each metric space S will be assigned an element of the set $\{-1, 0, 1, 2, \cdots, \infty\}$, called the large inductive dimension of S, written Ind S. To begin, Ind $\varnothing = -1$. Next, if k is a nonnegative integer, we will say that Ind $S \leq k$ iff any two disjoint closed sets in S can be separated by a set L with Ind $L \leq k - 1$. We write Ind $S = k$ iff Ind $S \leq k$ but Ind $S \not\leq k - 1$. We write Ind $S = \infty$ iff Ind $S \leq k$ is false for all integers k.

In general metric spaces, the large and small inductive dimension are not necessarily equal. But in separable spaces they are equal. This will be proved below (Corollary 3.2.15.) The first step in the proof is Theorem 3.2.4, which shows that if S is a separable metric space, then ind $S = 0$ if and only if Ind $S = 0$.

Before we proceed to higher dimensions, we prove a technical lemma, and discuss "sum theorems".

(3.2.6) LEMMA. *Let S be a metric space, let A and B be disjoint closed sets, and let $T \subseteq S$. (a) Let U and V be open sets with $U \supseteq A$, $V \supseteq B$, and $\overline{U} \cap \overline{V} = \varnothing$. If $L' \subseteq T$ separates (in T) the sets $T \cap \overline{U}$ and $T \cap \overline{V}$, then there is a set $L \subseteq S$ that separates A and B in S with $L \cap T \subseteq L'$. (b) Suppose, in addition, that T is closed. Then, for any set $L' \subseteq T$ separating $T \cap A$ and $T \cap B$ in T, there is $L \subseteq S$ separating A and B in S with $L \cap T \subseteq L'$.*

PROOF: (a) Since L' separates $T \cap \overline{U}$ and $T \cap \overline{V}$, we have $T \setminus L' = U' \cup V'$, with U' and V' open in T, $T \cap \overline{U} \subseteq U'$, and $T \cap \overline{V} \subseteq V'$. Now I claim that $A \cap \overline{V'} = \varnothing$. Indeed, $U \cap V' = T \cap U \cap V' \subseteq U' \cap V' = \varnothing$, and U is open, so $U \cap \overline{V'} = \varnothing$ and thus $A \cap \overline{V'} = \varnothing$. Similarly, $B \cap \overline{U'} = \varnothing$. Now U' and V' are disjoint and open in $U' \cup V'$, so $U' \cap \overline{V'} = \varnothing$ and $\overline{U'} \cap V' = \varnothing$. But then

$$(A \cup U') \cap \overline{(B \cup V')} = (A \cup U') \cap (\overline{B} \cup \overline{V'})$$
$$= (A \cap B) \cup (A \cap \overline{V'}) \cup (U' \cap B) \cup (U' \cap \overline{V'})$$
$$= \varnothing,$$

and similarly $\overline{(A \cup U')} \cap (B \cup V') = \varnothing$. So (by 3.2.3(1)) there exist disjoint open sets U'' and V'' with $A \cup U' \subseteq U''$ and $B \cup V' \subseteq V''$. Then let $L = S \setminus (U'' \cup V'')$. It separates A and B. Also

$$T \cap L = T \setminus (U'' \cup V'')$$
$$\subseteq T \setminus (U' \cup V')$$
$$= L'.$$

(b) Since L' separates $T \cap A$ and $T \cap B$ in T, there are disjoint sets U' and V', open in T, with $T \cap A \subseteq U'$, $T \cap B \subseteq V'$, and $T \setminus (U' \cup V') = L'$. Now the sets A and $(T \setminus U') \cup B$ are disjoint and closed, so there exists an open set U'' with

$$A \subseteq U'' \subseteq \overline{U''} \subseteq S \setminus ((T \setminus U') \cup B).$$

Similarly, B and $(T \setminus V') \cup \overline{U''}$ are disjoint and closed, so there exists an open set V'' with

$$B \subseteq V'' \subseteq \overline{V''} \subseteq S \setminus ((T \setminus V') \cup \overline{U''}).$$

So $A \subseteq U''$, $B \subseteq V''$, and $\overline{U''} \cap \overline{V''} = \varnothing$. Then apply part (a). ☺

This enables us to prove a generalization of Theorem 3.2.4:

(3.2.7) COROLLARY. *Let S be a separable metric space, let A and B be disjoint closed sets in S, and let $T \subseteq S$ be a subset with ind $T = 0$. Then there is a set L that separates A and B in S such that $L \cap T = \varnothing$.*

PROOF: Choose U and V open sets with $A \subseteq U$, $B \subseteq V$, and $\overline{U} \cap \overline{V} = \varnothing$. The sets $\overline{U} \cap T$ and $\overline{V} \cap T$ are disjoint and closed in T, which is zero-dimensional, so they can be separated by $L' = \varnothing$. Apply Lemma 3.2.6 to get L separating A and B with $L \cap T \subseteq L' = \varnothing$. ☺

Sum theorems. Before we proceed with the separation theorems for higher dimensions, we will discuss "sum theorems". These results deal with the dimension of a union of sets.

It is not true in general that the union of two zero-dimensional sets is zero-dimensional. Indeed, the set of rational numbers is zero-dimensional and the set of irrational numbers is zero-dimensional, but their union is the whole line, and has small inductive dimension 1. But it is true that the union of two *closed* zero-dimensional sets is zero-dimensional. In fact, the same is true for the union of countably many closed zero-dimensional sets:

(3.2.8) THEOREM. *Let S be a metric space, and suppose $T_n \subseteq S$ is a closed set with $\operatorname{Ind} T_n = 0$ for $n = 1, 2, \cdots$. Then $\operatorname{Ind} \bigcup_{n \in \mathbb{N}} T_n = 0$.*

PROOF: Write $T = \bigcup_{n \in \mathbb{N}} T_n$. Let E and F be disjoint closed sets in T. I must find a clopen set U with $E \subseteq U$ and $F \cap U = \varnothing$.

First, since $\operatorname{Ind} T_1 = 0$, there is a set A_1, clopen in T_1, with $T_1 \cap E \subseteq A_1$ and $T_1 \cap F \subseteq T_1 \setminus A_1$. Then the sets $E \cup A_1$ and $F \cup (T_1 \setminus A_1)$ are disjoint and closed. So there exist sets U_1 and V_1, open in S, with $E \cup A_1 \subseteq U_1$, $E \cup (T_1 \setminus A_1) \subseteq V_1$, and $\overline{U_1} \cap \overline{V_1} = \varnothing$.

Next, since $\operatorname{Ind} T_2 = 0$, there is a set A_2, clopen in T_2, with $T_2 \cap \overline{U_1} \subseteq A_2$ and $T_2 \cap \overline{V_1} \subseteq T_2 \setminus A_2$. Then the sets $\overline{U_1} \cup A_2$ and $\overline{V_1} \cup (T_2 \setminus A_2)$ are disjoint and closed. So there exist sets U_2 and V_2, open in S, with $\overline{U_1} \cup A_2 \subseteq U_2$, $\overline{V_1} \cup (T_2 \setminus A_2) \subseteq V_2$, and $\overline{U_2} \cap \overline{V_2} = \varnothing$.

Continue. We get in this way sequences (U_n) and (V_n). Then write $U = \bigcup_{n \in \mathbb{N}} U_n$ and $V = \bigcup_{n \in \mathbb{N}} V_n$. Then U and V are disjoint open sets with $U \supseteq E$ and $V \supseteq F$. Also, $U \cup V \supseteq \bigcup_{n \in \mathbb{N}} T_n = T$, so U and V are clopen in T. ☺

Of course, the same result is true for the small inductive dimension in a separable space, by Theorem 3.2.4.

(3.2.9) COROLLARY. *Let S be a separable metric space, and suppose $T_n \subseteq S$ is a closed set with $\operatorname{ind} T_n = 0$ for $n = 1, 2, \cdots$. Then $\operatorname{ind} \bigcup_{n \in \mathbb{N}} T_n = 0$.*

The sum theorem for dimensions other than 0 is discussed below: Theorem 3.2.12.

Here is what we can say when the sets are not closed:

(3.2.10) THEOREM. *Let S be a separable metric space, and let $A, B \subseteq S$. Then $\operatorname{ind}(A \cup B) \leq 1 + \operatorname{ind} A + \operatorname{ind} B$.*

PROOF: If either $\operatorname{ind} A = \infty$ or $\operatorname{ind} B = \infty$, then the inequality is clear. So suppose they are both finite, say $\operatorname{ind} A = m$, $\operatorname{ind} B = n$. The proof proceeds by induction on the sum $m + n$.

For $n + m = -2$, both A and B are empty, so $A \cup B$ is also empty, and $\operatorname{ind}(A \cup B) = -1 = 1 + (-1) + (-1)$.

So assume that the result is known for smaller sums. Let $x \in A \cup B$. Then either $x \in A$ or $x \in B$. Suppose $x \in A$. Let V be open in $A \cup B$ with $x \in V$. Then the two sets $\{x\}$ and $A \setminus V$ are separated in A by a set L' with ind $L' \le m - 1$. So by Lemma 3.2.6, there is a set L that separates $\{x\}$ and $(A \cup B) \setminus V$ in $A \cup B$, and $L \cap A \subseteq L'$. Now $L = (L \cap A) \cup (L \cap B)$, so by the induction hypothesis, ind $L \le 1 + (m - 1) + n = m + n$. Therefore I have shown that ind $(A \cup B) \le m + n + 1$, as required. ☺

(3.2.11) COROLLARY. *The union of $n + 1$ zero-dimensional sets in a separable metric space has small inductive dimension $\le n$.*

Next we prove the sum theorem for higher dimensions:

(3.2.12) THEOREM. *Let S be a separable metric space, and let closed sets $T_n \subseteq S$ satisfy* ind $T_n \le k$ *for $n = 1, 2, \cdots$. Then* ind $\bigcup_{n \in \mathbb{N}} T_n \le k$.

PROOF: Write $T = \bigcup_{n \in \mathbb{N}} T_n$. The result is clearly true for $k = \infty$. So suppose k is finite. The proof is by induction on k. The case $k = 0$ has been done already. Suppose $k \ge 1$, and assume the result is known for smaller values.

For each n, let $\mathcal{B}_n$ be a base for the open sets of T_n consisting of sets with boundary of dimension $< k$. By the Lindelöf property, we may assume that the bases $\mathcal{B}_n$ are countable. For all n and all $U \in \mathcal{B}_n$, we have ind $\partial_{T_n} U \le k - 1$. By the induction hypothesis, the countable union

$$Y = \bigcup_{n \in \mathbb{N}} \bigcup_{U \in \mathcal{B}_n} (\partial_{T_n} U)$$

also has ind $Y \le k - 1$. But the space $Z_n = T_n \setminus Y$ has the family

$$\{ U \setminus Y : U \in \mathcal{B}_n \}$$

as a base for its open sets, and the sets $U \setminus Y$ are clopen in Z_n. So ind $Z_n \le 0$. Now consider the union

$$Z = \bigcup_{n \in \mathbb{N}} Z_n.$$

Each $Z_n = T_n \setminus Y = T_n \cap Z$ is closed in Z, so by Theorem 3.2.12, ind $Z \le 0$. Thus (by Theorem 3.2.10)

$$\text{ind } T = \text{ind} (Y \cup Z) \le 1 + (k - 1) + 0 = k. \; ☺$$

Examination of the proof yields the converse of Corollary 3.2.11:

(3.2.13) COROLLARY. *Let S be a separable metric space with finite small inductive dimension k. Then S is the union of $k + 1$ zero-dimensional sets.*

More separation. Now we return to the problem of separation.

(3.2.14) THEOREM. *Let S be a separable metric space, let A and B be disjoint closed subsets. Let $k \geq 0$ be an integer, and let $T \subseteq S$ have small inductive dimension k. Then there is a set L separating A and B with $\mathrm{ind}\,(T \cap L) \leq k - 1$.*

PROOF: The case $k = 0$ has already been done (Corollary 3.2.7). So suppose $k \geq 1$. By Corollary 3.2.13, we can write $T = Y \cup Z$, with $\mathrm{ind}\,Y = k - 1$ and $\mathrm{ind}\,Z = 0$. Now (again by Corollary 3.2.7) there is a set L separating A and B with $L \cap Z = \varnothing$. Then $L \cap T \subseteq Y$. So $\mathrm{ind}\,(L \cap T) \leq \mathrm{ind}\,Y \leq k - 1$. ☺

Next is the last step in the proof of the equivalence of the large and small inductive dimensions.

(3.2.15) COROLLARY. *Let S be a separable metric space. Then $\mathrm{ind}\,S = \mathrm{Ind}\,S$.*

PROOF: It is clear that $\mathrm{ind}\,S \leq \mathrm{Ind}\,S$. So we must prove $\mathrm{Ind}\,S \leq \mathrm{ind}\,S$. This is clear if $\mathrm{ind}\,S = \infty$. So suppose $\mathrm{ind}\,S = k$ is finite. The case $k = 0$ is Theorem 3.2.4. Suppose $k \geq 1$ and the result is known for smaller values of the dimension. Let A and B be disjoint closed sets in S. Then by Theorem 3.2.14, A and B can be separated by a set L with $\mathrm{ind}\,L \leq k - 1$. But by the induction hypothesis, $\mathrm{Ind}\,L \leq k - 1$. Therefore we have $\mathrm{Ind}\,S \leq k$. ☺

Another consequence is this separation theorem, which generalizes Theorem 3.2.4 in another way:

(3.2.16) COROLLARY. *Suppose S is a separable metric space with $\mathrm{ind}\,S \leq n - 1$. Let $A_1, B_1, A_2, B_2, \cdots, A_n, B_n$ be $2n$ closed sets in S with $A_i \cap B_i = \varnothing$ for all i. Then there exist sets $L_1, L_2, \cdots, L_n$ such that L_i separates A_i and B_i for all i, and the intersection $\bigcap_{i=1}^{n} L_i = \varnothing$.*

PROOF: First, there is a set L_1 that separates A_1 and B_1, such that $\mathrm{ind}\,L_1 \leq n - 2$. Applying the theorem with $T = L_1$, we get a set L_2 that separates A_2 and B_2 with $\mathrm{ind}\,(L_1 \cap L_2) \leq n - 3$. Continuing in this way, we get $L_1, L_2, \cdots, L_n$ with $\mathrm{ind}\,(L_1 \cap L_2 \cap \cdots \cap L_n) \leq -1$, so it is empty. ☺

*3.3. TWO-DIMENSIONAL EUCLIDEAN SPACE

In this section we will prove that the topological dimension of the plane $\mathbf{R}^2$ is really 2.

We will use this notation:

$$B = \left\{\, x \in \mathbf{R}^2 : |x| \leq 1 \,\right\}, \qquad \text{the unit disk;}$$
$$T = \left\{\, x \in \mathbf{R}^2 : |x| = 1 \,\right\}, \qquad \text{the unit circle;}$$
$$S = [-1, 1] \times [-1, 1], \qquad \text{a square.}$$

Degree (mod 2). The preliminaries to the discussion of the dimension of the plane involve a "homological" discussion of the degree of a map of the circle to itself, and the Brouwer fixed point theorem.

---

*An optional section.

(3.3.1) THEOREM. *There is no continuous function* $f : B \to T$ *with* $f(x) = x$ *for all* $x \in T$.

PROOF: If n is a positive integer, we define the **subdivision** $A(n)$ of the circle to be the set $\{a_0, a_1, \cdots, a_{n-1}\}$ of points on the circle T, starting with $a_0 = (1, 0)$, and continuing counterclockwise around the circle with equal spacing. So

$$a_j = \left(\cos \frac{2\pi j}{n}, \sin \frac{2\pi j}{n} \right).$$

When n is large, consecutive points a_{j-1} and a_j are close together. Sometimes we may write $a_n = a_0$.

Given a continuous function $g : T \to T$, consider the image points $g(a_0)$, $g(a_1)$, $\cdots, g(a_n)$. They may not occur in order around the boundary of the circle. Assuming that two consecutive points $g(a_{j-1})$ and $g(a_j)$ are close together (say, no more than $1/4$ of the circumference apart), let U_j be the shorter of the two arcs of the circle with endpoints $g(a_{j-1})$ and $g(a_j)$. If $g(a_{j-1}) = g(a_j)$, then U_j is just that single point. Now if $y \in T$ is not one of the points $g(a_j)$, we let $N(A(n), g, y)$ denote the number of the intervals U_j that y is in. Let $\widetilde{N}(A(n), g, y)$ be the residue* modulo 2 of $N(A(n), g, y)$.

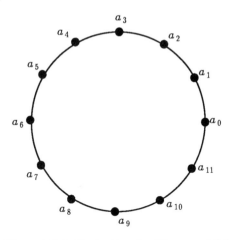

Figure 3.3.1. Illustration for Theorem 3.3.1.

I claim that the number $\widetilde{N}(A(n), g, y)$ is independent of the choice of y, as long as it is not one of the points $g(a_j)$. Indeed, think of moving y around the circle. As long as we do not cross one of the points $g(a_j)$, the number $N(A(n), g, y)$ remains unchanged. When we do cross one of the points $g(a_j)$, the

---

*There are only two possible values: 0 represents "even" and 1 represents "odd".

number $N(A(n), g, y)$ may remain unchanged (if the two arcs U_j and U_{j+1} are on opposite sides of $g(a_j)$) or it may increase or decrease by 2 (if the two arcs U_j and U_{j+1} are on the same side of $g(a_j)$). It any case, the parity $\tilde{N}(A(n), g, y)$ remains unchanged.

Therefore we will write $\tilde{N}(A(n), g)$ for the value of all $\tilde{N}(A(n), g, y)$.

Now let $f: B \to T$ be continuous, and suppose that $f(x) = x$ for all $x \in T$. For $0 \le r \le 1$, let

$$g_r(x) = f(rx) \qquad \text{for } x \in T.$$

Since B is compact, the function f is uniformly continuous. Therefore there is $\delta > 0$ so that $|x - y| < \delta$ implies $|f(x) - f(y)| < 1$. If n is large enough, consecutive points a_j of the subdivision $A(n)$ are within distance δ of each other. For this value of n, all of the functions g_r satisfy the assumption required above that consecutive points $g_r(a_{j-1})$, $g_r(a_j)$ have distance less than $1/4$ of the circumference of the circle. Fix this value of n.

Consider the numbers $N(A(n), g_r)$. The function g_1 is the identity on T, so clearly $N(A(n), g_1) = 1$. The function g_0 maps everything to the single point $f(0)$, so $N(A(n), g_0) = 0$. Now I claim that both of the sets $\{ r \in [0, 1] : \tilde{N}(A(n), g_r) = 0 \}$ and $\{ r \in [0, 1] : \tilde{N}(A(n), g_r) = 1 \}$ are open in $[0, 1]$. Fix some value $r_0 \in \{ r \in [0, 1] : \tilde{N}(A(n), g_r) = 0 \}$. Choose a point $y \in T$ not equal to any of the points $g_{r_0}(a_j)$. There is a minimum distance ε from y to the set $\{ g_{r_0}(a_0), g_{r_0}(a_1), \cdots, g_{r_0}(a_{n-1}) \}$. By the uniform continuity of f, there is $\delta > 0$ so that if $|r - r_0| < \delta$, then $|g_r(a_j) - g_{r_0}(a_j)| < \varepsilon$ for all j. This means that y lies in exactly the same arcs defined by g_r as arcs defined by g_{r_0}. So $N(A(n), g_r) = N(A(n), g_{r_0})$. Therefore $B_\delta(r_0)$ lies entirely in the set $\{ r \in [0, 1] : \tilde{N}(A(n), g_r) = 0 \}$. So it is an open set.

Thus $[0, 1]$ is the union of two disjoint nonempty subsets, both open in $[0, 1]$. This is impossible by Corollary 3.1.2. So the assumption that the function f exists is not tenable. ☺

An easy consequence is Brouwer's fixed point theorem (for the disk). We will use a calculation, which is left to the reader. You can think of it as a test of your analytic geometry and algebra skills.

(3.3.2) EXERCISE. *Let (x, y) and (a, b) be distinct points in the disk B. Then the point (u, v) where the ray from (a, b) through (x, y) intersects the circle is given by $u = x + t(x - a)$ and $v = y + t(y - b)$, where*

$$t = \frac{-(x - a)x - (y - b)y + \sqrt{(x - a)^2 + (y - b)^2 - (ay + bx)^2}}{(x - a)^2 + (y - b)^2}.$$

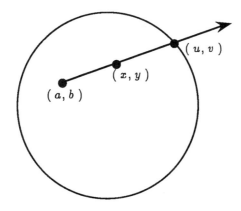

Figure 3.3.2. Illustration for Exercise 3.3.2.

(3.3.3) BROUWER'S FIXED POINT THEOREM. *If* $g : B \rightarrow B$ *is continuous, then there is a point* $x \in B$ *with* $g(x) = x$.

PROOF: Suppose there is no such fixed point. Define a function f as follows. If $x \in B$, then $g(x) \neq x$ is also in B; let $f(x)$ be the point where the ray from $g(x)$ through x intersects the circle T. By Exercise 3.3.2, we can see that the function f is continuous. Also, $f(x) = x$ if $x \in T$. This contradicts Theorem 3.3.1. So in fact g has a fixed point. ☺

Note that the square S is homeomorphic to the disk B, so Brouwer's fixed point theorem also holds for the square.

Topological dimension of the plane. Now I will show that the plane $\mathbf{R}^2$ has ind $\mathbf{R}^2 = 2$. The usual base for the topology of $\mathbf{R}^2$, consists of the open balls. Their boundaries are circles, which have small inductive dimension 1. So clearly ind $\mathbf{R}^2 \leq 2$. I will show that the square S has ind ≥ 2. Since S is a subset of $\mathbf{R}^2$, it follows that ind $\mathbf{R}^2 \geq 2$. Therefore ind $\mathbf{R}^2 = 2$.

(3.3.4) THEOREM. *The square* $S = [-1, 1] \times [-1, 1]$ *has small inductive dimension* ≥ 2.

PROOF: The "left" and "right" sides

$$A_1 = \{-1\} \times [-1, 1], \qquad B_1 = \{1\} \times [-1, 1],$$

are disjoint closed sets. The "bottom" and "top" sides

$$A_2 = [-1, 1] \times \{-1\}, \qquad B_2 = [-1, 1] \times \{1\},$$

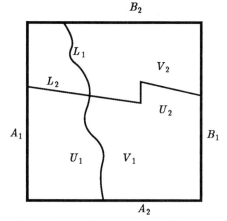

Figure 3.3.4. Illustration for Theorem 3.3.4.

are disjoint closed sets. Suppose L_1 separates A_1 and B_1; and L_2 separates A_2 and B_2. I will show that $L_1 \cap L_2 \neq \varnothing$. Therefore, by Corollary 3.2.16, we will have ind $S \geq 2$.

Write $S \setminus L_1 = U_1 \cup V_1$, where U_1 and V_1 are disjoint open sets, $A_1 \subseteq U_1$, $B_1 \subseteq V_1$. Similarly $S \setminus L_2 = U_2 \cup V_2$, where U_2 and V_2 are disjoint open sets, $A_2 \subseteq U_2$, $B_2 \subseteq V_2$.

Define a function $g_1 \colon S \to \mathbf{R}$ as follows: If $x = (x_1, x_2)$, then

$$g_1(x) = \begin{cases} x_1 + \operatorname{dist}(x, L_1) & \text{if } x \in U_1 \\ x_1 & \text{if } x \in L_1 \\ x_1 - \operatorname{dist}(x, L_1) & \text{if } x \in V_1. \end{cases}$$

Then g_1 is continuous, since it satisfies $|g_1(x) - g_1(y)| \leq 2|x - y|$ for $x, y \in S$. Also, I claim that $-1 \leq g_1(x) \leq 1$. To see this, consider the case $x \in U_1$. (The case $x \in V_1$ is similar, and the case $x \in L_1$ is trivial.) Write $x = (x_1, x_2)$. First, $g_1(x) \geq x_2 \geq -1$. Consider the horizontal line segment

$$Q = [x_1, 1] \times \{x_2\}$$

from x to the right edge of the square. The left endpoint x is in the open set U_1; the right endpoint $(1, x_2)$ is in the set B_1, so it is in the open set V_1. By Corollary 3.1.2, there is some point (z_1, x_2) between them that is a boundary point in Q of $U_1 \cap Q$. Therefore (z_1, x_2) belongs to L_1. Then $\operatorname{dist}(x, L_1) \leq |z_1 - x_1|$, so $g_1(x) \leq x_1 + (z_1 - x_1) = z_1 \leq 1$.

Define similarly a function $g_2 \colon S \to \mathbf{R}$ as follows: If $x = (x_1, x_2)$, then

$$g_2(x) = \begin{cases} x_2 + \operatorname{dist}(x, L_2) & \text{if } x \in U_2 \\ x_2 & \text{if } x \in L_2 \\ x_2 - \operatorname{dist}(x, L_2) & \text{if } x \in V_2. \end{cases}$$

Then g_2 is continuous, and $-1 \leq g_2(x) \leq 1$ for $x \in S$.

So the function g defined by $g(x) = (g_1(x), g_2(x))$ is continuous and maps S into S. By Brouwer's fixed point theorem, the map g has a fixed point. If $x = (x_1, x_2)$ is a fixed point, then $g_1(x) = x_1$, so $\text{dist}(x, L_1) = 0$. But L_1 is closed, so $x \in L_1$. Similarly $g_2(x) = x_2$, so $x \in L_2$. Therefore $L_1 \cap L_2 \neq \varnothing$. ☺

In fact, it is known that ind $\mathbf{R}^d = d$ for all positive integers d. For a proof, along the same lines as used in this section, see [30, Chapter IV].

*3.4. OTHER TOPOLOGICAL DIMENSIONS

Covering dimension. Figure 3.4.1 suggests that if a set should be considered 1-dimensional, then it can be covered by small open sets that intersect only 2 at a time. (That is: any 3 of the sets have empty intersection; or, each point belongs to at most 2 of the sets.) A set should be considered 0-dimensional if it can be covered by small open sets that are disjoint. So perhaps a set is 2-dimensional if it can be covered by small open sets that intersect only 3 at a time. This idea is reasonable for compact metric spaces (see Theorem 3.4.6), but needs a bit of fine-tuning for non-compact spaces.

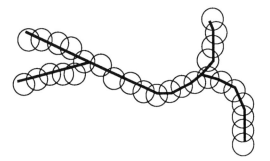

Figure 3.4.1. A 1-dimensional space.

If $\mathcal{A}$ and $\mathcal{B}$ are two covers of a metric space S, we say that $\mathcal{B}$ is a **refinement** of $\mathcal{A}$ iff for every $B \in \mathcal{B}$ there is $A \in \mathcal{A}$ with $B \subseteq A$. We also say $\mathcal{B}$ **refines** $\mathcal{A}$. For example, a subcover of $\mathcal{A}$ is a refinement of $\mathcal{A}$.

(3.4.2) EXERCISE. *Let S be a metric space. Then S is compact if and only if each open cover is refined by some finite open cover.*

The **order** of a family $\mathcal{A}$ of sets is $\leq n$ iff any $n + 2$ of the sets have empty intersection. It has order n iff it has order $\leq n$ but does not have order $\leq n - 1$.

---

*An optional section.

For example, a family $\mathcal{A}$ of nonempty sets is disjoint if and only if its order is 0. A family of sets $\mathcal{A}$ has order -1 iff it is the singleton $\{\varnothing\}$. Now

$$\mathbb{R} = \bigcup_{n \in \mathbb{Z}} (n-1, n+1)$$

exhibits a cover of $\mathbb{R}$ by open intervals, with order 1. The point of the covering dimension is that there is no cover of $\mathbb{R}^2$ by bounded open sets with order 1.

Let S be a metric space. Let $n \geq -1$ be an integer. We say that S has **covering dimension** $\leq n$ iff every finite open cover of S has an open refinement with order $\leq n$. The covering dimension is n iff the covering dimension is $\leq n$ but not $\leq n-1$. We will write Cov $S = n$ in that case. If the covering dimension is $\leq n$ for no integer n, then we say Cov $S = \infty$.

The first few cases are easy to relate to what we have already discussed. First, if Cov $S = -1$ then the open cover $\{S\}$ is refined by a cover of order -1, which is necessarily $\{\varnothing\}$; this is a cover only if $S = \varnothing$. So Cov $S = -1$ if and only if $S = \varnothing$.

(3.4.3) THEOREM. *Let S be a metric space, and let n be a nonnegative integer. The following are equivalent:*

(1) Cov $S \leq n$

(2) *If $\{U_1, U_2, \cdots, U_k\}$ is any finite open cover of S, then there exist open sets $B_1 \subseteq U_1$, $B_2 \subseteq U_2, \cdots$, $B_k \subseteq U_k$, such that $\{B_1, B_2, \cdots, B_k\}$ is an open cover of S with order $\leq n$.*

(3) *If $\{U_1, U_2, \cdots, U_{n+2}\}$ is an open cover of S, then there exist open sets $B_1 \subseteq U_1$, $B_2 \subseteq U_2, \cdots$, $B_{n+2} \subseteq U_{n+2}$, such that $\bigcup_{i=1}^{n+2} B_i = S$ and $\bigcap_{i=1}^{n+2} B_i = \varnothing$.*

(4) *If $\{U_1, U_2, \cdots, U_{n+2}\}$ is an open cover of S, then there exist closed sets $F_1 \subseteq U_1$, $F_2 \subseteq U_2, \cdots$, $F_{n+2} \subseteq U_{n+2}$, such that $\bigcup_{i=1}^{n+2} F_i = S$ and $\bigcap_{i=1}^{n+2} F_i = \varnothing$.*

PROOF: (2) $\Longrightarrow$ (3) and (2) $\Longrightarrow$ (1) are obvious.

(1) $\Longrightarrow$ (2). Suppose Cov $S \leq n$. The open cover $\{U_1, U_2, \cdots, U_k\}$ admits an open refinement $\mathcal{W}$ of order $\leq n$. For each $W \in \mathcal{W}$ there is at least one i such that $W \subseteq U_i$; choose one of them, and call it $i(W)$. Now for each i, let

$$B_i = \bigcup \{ W \in \mathcal{W} : i(W) = i \}.$$

Then the sets B_i are open, and $\bigcup_i B_i = \bigcup_{W \in \mathcal{W}} W = S$. If $x \in S$, then (since $\mathcal{W}$ has order $\leq n$) it belongs to at most $n+1$ of the sets W. But $x \in B_i$ only if $x \in W$ for some W with $i(W) = i$. So x belongs to at most $n+1$ of the sets B_i.

(3) $\Longrightarrow$ (2). Suppose S has the property (3). Let $\{U_1, U_2, \cdots, U_k\}$ be an open cover of S. If $k \leq n+1$, then this cover itself already has order $\leq n$. So suppose $k \geq n+2$. Now write $W_1 = U_1$, $W_2 = U_2, \cdots$, $W_{n+1} = U_{n+1}$ and

$W_{n+2} = \bigcup_{i=n+2}^{k} U_i$. Then these sets cover S, so by the hypothesis (3), there exist open sets $V_i \subseteq W_i$ with $\bigcup_{i=1}^{n+2} V_i = S$ and $\bigcap_{i=1}^{n+2} V_i = \varnothing$. Let $B_i = V_i$ for $i \leq n+1$ and $B_i = V_{n+2} \cap U_i$ for $i \geq n+2$. Then $B_i \subseteq U_i$ for all i, we have $\bigcup_{i=1}^{k} B_i = S$, and $\bigcap_{i=1}^{n+2} B_i = \varnothing$. Repeat this construction a finite number of times, once for each subset of $\{1, 2, \cdots, k\}$ with $n+2$ elements, to arrange the same conclusion with *all* of the intersections of size $n+2$ empty.

(3) $\Longrightarrow$ (4). There exist open sets $B_i \subseteq U_i$ with $\bigcup B_i = S$ and $\bigcap B_i = \varnothing$. Now $S \setminus B_1 \subseteq \bigcup_{i=2}^{n+2} B_i$, so by Corollary 3.2.2, there is an open set V_1 with $S \setminus B_1 \subseteq V_1 \subseteq \overline{V_1} \subseteq \bigcup_{i=2}^{n+2} B_i$. Let $F_1 = S \setminus V_1$. So we have $F_1 \subseteq B_1$ and $F \cup \bigcup_{i=2}^{n} B_i = S$. Next, there is an open set V_2 with $S \setminus B_2 \subseteq V_2 \subseteq \overline{V_2} \subseteq (S \setminus \overline{V_1}) \cup \bigcup_{i=2}^{n+2} B_i$. Let $F_2 = S \setminus V_2$. So we have $F_2 \subseteq B_2$ and $F_1 \cup F_2 \cup \bigcup_{i=3}^{n+2} B_i = S$. Continue in this way.

(4) $\Longrightarrow$ (3). There exist closed sets F_i as in (4). Now the closed set F_1 is a subset of the open set $U_1 \cap (S \setminus \bigcap_{i=2}^{n+2} F_i)$, so there is an open set B_1 with $F_2 \subseteq B_1 \subseteq \overline{B_1} \subseteq U_1 \cap (S \setminus \bigcap_{i=2}^{n+2} F_i)$. So $\overline{B_1} \subseteq U_1$ and $\overline{B_1} \cap \bigcap_{i=2}^{n+2} F_i = \varnothing$. Next, there is an open set B_2 with $F_2 \subseteq B_2 \subseteq \overline{B_2} \subseteq U_2 \cap (S \setminus (\overline{B_1} \cap \bigcap_{i=3}^{n+2} F_i))$, so $\overline{B_2} \subseteq U_2$ and $\overline{B_1} \cap \overline{B_2} \cap \bigcap_{i=3}^{n+2} F_i = \varnothing$. Continue in this way. ☺

(3.4.4) THEOREM. *Let S be a metric space. Then $\operatorname{Cov} S = 0$ if and only if $\operatorname{Ind} S = 0$.*

PROOF: Suppose $\operatorname{Ind} S = 0$. Let U_1, U_2 be open sets with $U_1 \cup U_2 = S$. Then their complements $F_1 = S \setminus U_1$ and $F_2 = S \setminus U_2$ are disjoint closed sets. Therefore, there exists a clopen set V with $F_1 \subseteq V$ and $F_2 \cap V = \varnothing$. Thus the sets $B_1 = S \setminus V$ and $B_2 = V$ satisfy $B_i \subseteq U_i$, $B_1 \cup B_2 = S$, and $B_1 \cap B_2 = \varnothing$. By Theorem 3.4.3, we have $\operatorname{Cov} S \leq 0$. But $S \neq \varnothing$, so $\operatorname{Cov} S = 0$.

Conversely, suppose $\operatorname{Cov} S = 0$. Let F_1 and F_2 be disjoint closed sets. Then their complements $U_1 = S \setminus F_1$ and $U_2 = S \setminus F_2$ constitute an open cover of S. Therefore, by Theorem 3.4.2, there exist open sets $B_i \subseteq U_i$ with $B_1 \cup B_2 = S$ and $B_1 \cap B_2 = \varnothing$. Thus $V = B_2$ is clopen, $V \supseteq F_1$, and $V \cap F_2 = \varnothing$. This shows that $\operatorname{Ind} S = 0$. ☺

In separable metric spaces S, it is true in general that $\operatorname{Cov} S = \operatorname{Ind} S = \operatorname{ind} S$. (See, for example, [30, Chapter V] or [18, Theorem 1.7.9].) I will prove only part of this result. First I prove the easy direction:

(3.4.5) THEOREM. *Let S be a separable metric space. Then $\operatorname{Cov} S \leq \operatorname{ind} S$.*

PROOF: The result is clearly true if $\operatorname{ind} S = \infty$. So suppose we have $\operatorname{ind} S < \infty$. Let $n = \operatorname{ind} S$. I must show $\operatorname{Cov} S \leq n$. Now by Corollary 3.2.13, there exist zero-dimensional sets $Y_1, Y_2, \cdots, Y_{n+1}$ with $S = \bigcup_{j=1}^{n+1} Y_j$. Let $\{U_1, U_2, \cdots, U_k\}$ be a finite open cover of S. Then for each j, the family $\{U_1 \cap Y_j, U_2 \cap Y_j, \cdots, U_k \cap Y_j\}$ is a finite open cover of Y_j. Now $\operatorname{ind} Y_j = 0$, so $\operatorname{Cov} Y_j = 0$, and therefore there exist disjoint open sets $B_{ij} \subseteq U_i$ with $\bigcup_{i=1}^{k} B_{ij} \supseteq Y_j$. Now the family

$$\mathcal{B} = \{ B_{ij} : 1 \leq i \leq k, 1 \leq j \leq n+1 \}$$

covers S and refines $\{U_1, U_2, \cdots, U_k\}$. If we have any $n + 2$ elements B_{ij} of $\mathcal{B}$, some two of them have the same second index j, so the intersection of the $n + 2$ sets is empty. Therefore $\mathcal{B}$ has order $\leq n$. This completes the proof that Cov $S \leq n$. ☺

I will prove the other direction only for compact spaces S. If S is compact, there is a simpler description of the covering dimension. If $\mathcal{A}$ is a cover, then its **mesh** is $\sup_{A \in \mathcal{A}} \operatorname{diam} A$.

(3.4.6) THEOREM. *Let S be a compact metric space, and let $n \geq -1$ be an integer. Then* Cov $S \leq n$ *if and only if for every $\varepsilon > 0$, there is an open cover of S with order $\leq n$ and mesh $\leq \varepsilon$.*

PROOF: Suppose Cov $S \leq n$. Let $\varepsilon > 0$. The collection $\mathcal{U}$ of all open sets of diameter $\leq \varepsilon$ is a cover of S. So it has a refinement $\mathcal{B}$ of order at most n. But $\mathcal{B}$ has mesh $\leq \varepsilon$.

Conversely, suppose that for every $\varepsilon > 0$, there is an open cover of S with order $\leq n$ and mesh $\leq \varepsilon$. Let $\mathcal{U}$ be any finite open cover of S. By Theorem 2.2.22, the Lebesgue number r of $\mathcal{U}$ is positive. Let $\mathcal{B}$ be an open cover of S with order at most n and mesh less than the minimum of r and ε. Then by the defining property of the Lebesgue number, $\mathcal{B}$ is a refinement of $\mathcal{U}$. So Cov $S \leq n$. ☺

The second part of the equality (for compact spaces) of the covering dimension with the other topological dimensions will be proved in a way that also establishes another useful characterization of topological dimension for compact spaces. A function $h: W \to S$ is said to be **at most n-to-one** iff for each $x \in S$ there are at most n points $y \in W$ with $h(y) = x$.

(3.4.7) THEOREM. *Let S be a compact metric space. Let $n \geq 0$ be an integer. Then the following are equivalent:*

1. ind $S \leq n$.
2. Cov $S \leq n$.
3. *There is a compact zero-dimensional metric space W and a continuous map h of W onto S that is at most $(n + 1)$-to-one.*

PROOF: (1) $\Longrightarrow$ (2) is Theorem 3.4.5.

(2) $\Longrightarrow$ (3). Begin with the cover $\mathcal{U}_0 = \{S\}$, and (recursively) choose finite open covers $\mathcal{U}_k$ of order $\leq n$ such that the mesh of $\mathcal{U}_k$ is less than half of the Lebesgue number of $\mathcal{U}_{k-1}$. We can use this sequence of covers to define a tree: the nodes of the kth generation are the elements of $\mathcal{U}_k$; for each $U \in \mathcal{U}_k$, there is $V \in \mathcal{U}_{k-1}$ with $\overline{U} \subseteq V$; let V be the parent of U. (If there is more than one such set V, choose one.) If we label the edges of the tree (an edge goes from the parent to the child) by a countable set E, then the set of nodes of the tree is in one-to-one correspondence with the set $E_z^{(*)}$ finite paths in the tree starting at the root z. Write U_α for the open set corresponding to the path $\alpha \in E_z^{(*)}$. Thus, the cover $\mathcal{U}_k$ is $\{U_\alpha : \alpha \in E^{(k)}\}$. Let $W = E_z^{(\omega)}$ be the set of infinite paths

starting at the root. As usual, W is an ultrametric space under the metric $\rho_{1/2}$, and therefore ind $W = 0$. Also W is compact, since each node has only finitely many children.

Define the map $h \colon W \to S$ as follows: If $\sigma \in E_z^{(\omega)}$ is an infinite path, the sets $\overline{U_{\sigma \restriction k}}$ are compact, $\overline{U_{\sigma \restriction 0}} \supseteq \overline{U_{\sigma \restriction 1}} \supseteq \cdots$, and $\lim_{k \to \infty}$ diam $\overline{U_{\sigma \restriction k}} = 0$. The set

$$\bigcap_{k \in \mathbb{N}} \overline{U_{\sigma \restriction k}} = \bigcap_{k \in \mathbb{N}} U_{\sigma \restriction k}$$

is nonempty (by compactness) and has diameter 0, so it consists of a single point. Let that point be $h(\sigma)$. Now $h\big[[\sigma \restriction k]\big] \subseteq U_{\sigma \restriction k}$, so h is continuous. Since the families $\mathcal{U}_k$ cover S, we may deduce that h is surjective.

Finally, I must show that the map is at most $(n+1)$-to-one. Suppose σ_1, σ_2, $\cdots$, $\sigma_{n+2} \in W$ are all different. Then there is a generation k so that

$$\sigma_1 \restriction k, \sigma_2 \restriction k, \cdots, \sigma_{n+2} \restriction k$$

are all different. But $\mathcal{U}_k$ has order at most n, so

$$\bigcap_{i=1}^{n+2} U_{\sigma_i \restriction k} = \varnothing.$$

Therefore not all of the points $h(\sigma_i)$ are equal.

(3) $\Longrightarrow$ (1). The proof is by induction on n. The case $n = 0$ and the induction step for $n \geq 1$ begin in the same way, however.

Let $x \in S$ and let $\varepsilon > 0$. I must show that there is an open set $V \subseteq S$ such that $x \in V \subseteq B_\varepsilon(x)$ and ind $\partial V \leq n - 1$. The set $D = h^{-1}\big[\{x\}\big]$ has at most $n + 1$ elements, say $D = \{z_1, z_2, \cdots, z_m\}$. The function h is continuous and W is zero-dimensional, so there exist clopen sets $U_i \subseteq W$ with $z_i \in U_i$ and $h[U_i] \subseteq B_\varepsilon(x)$. Let $F = \bigcup_{i=1}^m h[U_i]$. Since the sets U_i are clopen, the set F is closed. Let V be the interior of F, that is, the set of all the interior points of F. Then V is an open set.

First, I claim that $x \in V$, that is, x is an interior point of F. Suppose not. Then there is a sequence (x_k) in $S \setminus F$ with $x_k \to x$. Choose points $y_k \in W$ with $h(y_k) = x_k$. Taking a subsequence, we may assume that (y_k) converges. Its limit y satisfies $h(y) = x$, so in fact $y = z_i$ for some i. Now z_i is an interior point of U_i, so $y_k \in U_i$ for some k, and therefore $x_k = h(y_k) \in F$. This contradiction shows that x is an interior point of F.

Note that $\partial V \subseteq F \setminus V$. Now consider the subset

$$W_1 = h^{-1}[\partial V] \setminus (U_1 \cup U_2 \cup \cdots \cup U_m)$$

of W. It is closed, hence compact and zero-dimensional. I claim that h maps W_1 onto ∂V. Indeed, if $y \in \partial V$, then there is a sequence (y_k) in $S \setminus F$ with $y_k \to y$. Choose $w_k \in W$ with $h(w_k) = y_k$; by taking a subsequence, we may assume that

$w_k \to w$ for some $w \in W$. Of course, $h(w) = y$. Now $w_k \notin U_1 \cup U_2 \cup \cdots \cup U_m$, so $w \notin U_1 \cup U_2 \cup \cdots \cup U_m$, and therefore $y \in W_1$.

Next I claim that $\operatorname{ind} \partial V \leq n - 1$. We must distinguish the cases $n = 0$ and $n \geq 1$. First, suppose $n = 0$. Then h is one-to-one. Now $\partial V \subseteq F$ and $W_1 \subseteq h^{-1}[F] \setminus U_1 = \varnothing$. Therefore $\partial V = \varnothing$, or $\operatorname{ind} \partial V = -1 = n - 1$. Next suppose $n \geq 1$ and the result is known for smaller values of n. If $y \in \partial V$, then since $\partial V \subseteq F$, the map h sends at least one point of $U_1 \cup U_2 \cup \cdots \cup U_m$ onto y, so at most n points of W_1 are sent onto y. Thus the restriction $h \colon W_1 \to \partial V$ satisfies the hypotheses of the theorem for $n - 1$. By the induction hypothesis, $\operatorname{ind} \partial V \leq n - 1$.

This completes the proof that $\operatorname{ind} S \leq n$. ☺

(3.4.8) EXERCISE. *Compute the covering dimension of the line* $\mathbb{R}$.

Solution of systems of equations. One of the most important and useful topics of linear algebra is the solution of systems of linear equations. For example

$$
\begin{aligned}
x_1 + x_2 + x_3 &= 2 \\
2x_1 + 3x_2 - 4x_3 &= 4 \\
x_1 \quad\quad - x_3 &= -2
\end{aligned}
$$

has the solution $x_1 = -3/2$, $x_2 = 3$, $x_3 = 1/2$. (This is the only solution.) It is possible that a system has no solution:

$$
\begin{aligned}
x_1 + x_2 + x_3 &= 2 \\
2x_1 + 3x_2 - 4x_3 &= 4 \\
3x_1 + 4x_2 - 3x_3 &= 0
\end{aligned}
$$

It is possible that a system has infinitely many solutions:

$$
\begin{aligned}
x_1 + x_2 + x_3 &= 2 \\
2x_1 + 3x_2 - 4x_3 &= 4 \\
3x_1 + 4x_2 - 3x_3 &= 6
\end{aligned}
$$

Typically, if the number of variables exceeds the number of equations, then there are many solutions; if the number of equations exceeds the number of variables, then there are no solutions; and if the number of variables is the same as the number of equations, then there is exactly one solution.

These are only typical cases, however. It is possible for a system with more variables than equations to have no solutions:

$$
\begin{aligned}
x_1 + x_2 + x_3 &= 2 \\
x_1 + x_2 + x_3 &= 1
\end{aligned}
$$

And it is possible for a system with more equations than variables to have a solution:

$$x_1 + x_2 = 2$$
$$x_1 + 2x_2 = 3$$
$$x_1 - x_2 = 0$$

Here, there is a solution $x_1 = 1$, $x_2 = 1$. But it is not an essential solution: by that we mean that changing the equations by a tiny amount may yield a system with no solution:

$$x_1 + x_2 = 2.001$$
$$x_1 + 2x_2 = 2.999$$
$$x_1 - x_2 = 0$$

There is a similar situation for non-linear equations. A continuous equation in d variables is an equation of the form

$$f(x_1, x_2, \cdots, x_d) = 0,$$

where $f : \mathbf{R}^d \to \mathbf{R}$ is a continuous function. This will also be written $f(x) = 0$, and a solution is a point $x \in \mathbf{R}^d$. A system of m equations in d unknowns looks like

$$f_i(x) = 0 \qquad (i = 1, 2, \cdots, m),$$

where the $f_i : \mathbf{R}^d \to \mathbf{R}$ are given continuous functions. Typically, such a system has infinitely many solutions if $m < d$, no solutions if $m > d$, and a finite number of solutions if $m = d$. But of course, these typical results can (and often do) fail.

But there is still something that can be said about **essential solutions**. (The proofs require considerable knowledge of topological dimension theory.) If $m > d$, it is possible that the system $f_i(x) = 0$ $(i = 1, 2, \cdots, m)$ has a solution. But it has no essential solution. This means: for any $\varepsilon > 0$, there exists a system $g_i(x) = 0$ $(i = 1, 2, \cdots, m)$ with no solution, but $\rho_u(f_i, g_i) \le \varepsilon$ for all i. On the other hand, if $m \le d$, there exists a system $f_i(x) = 0$ $(i = 1, 2, \cdots, m)$ which has an essential solution; that is, for $\varepsilon > 0$ small enough, any system $g_i(x) = 0$ $(i = 1, 2, \cdots, m)$ has a solution provided that $\rho_u(g_i, f_i) \le \varepsilon$ for all i.

This may be used to define the dimension of an abstract metric space S. We simply replace the space $\mathbf{R}^d$ by the metric space S. Then we ask about systems of equations of the form $f_i(x) = 0$ $(i = 1, 2, \cdots, m)$, where $f_i : S \to \mathbf{R}$ are continuous functions. Whether or not the system has an essential solution may depend on the number of equations in the system, just as in the previous case. If S behaves like $\mathbf{R}^n$ in this regard, then we may consider S to have dimension n.

More precisely: the **equation dimension** of S is the largest integer n such that there exist n continuous functions $f_i \colon S \to \mathbf{R}$ ($i = 1, 2, \cdots, n$) for which the system $f_i(x) = 0$ ($i = 1, 2, \cdots, n$) has an essential solution. We may write eqd $S = n$.

In fact, for separable metric spaces, the equation dimension coincides with the topological dimensions already discussed (ind, Ind, Cov). I will illustrate here only the case $n = 0$, that is: eqd $S = 0$ if and only if Ind $S = 0$. For values other than 0, see [30, Chapter VI].

(3.4.9) THEOREM. *Let S be a metric space.* (a) *Suppose* Ind $S \leq 0$. *Let $f \colon S \to \mathbf{R}$ be a continuous function, and let $\varepsilon > 0$. Then there is a continuous function $g \colon S \to \mathbf{R}$ with $\rho_u(f, g) \leq \varepsilon$, but such that $g(x) \neq 0$ for all $x \in S$.* (b) *Suppose* Ind $S > 0$. *Then there is a continuous function $f \colon S \to \mathbf{R}$ such that $g(x) = 0$ has a solution for all continuous g with $\rho_u(f, g) \leq 1/2$.*

PROOF: (a) The two sets

$$A = \{\, x \in S : f(x) \geq \varepsilon/2 \,\},$$
$$B = \{\, x \in S : f(x) \leq -\varepsilon/2 \,\}$$

are closed and disjoint. Therefore there exists a clopen set U with $A \subseteq U$ and $U \cap B = \varnothing$. Define $g \colon S \to \mathbf{R}$ by

$$g(x) = \begin{cases} \max\{f(x), \varepsilon/2\} & \text{if } x \in U \\ \min\{f(x), -\varepsilon/2\} & \text{if } x \notin U. \end{cases}$$

Then g is continuous, since U is clopen. Now $g(x) = f(x)$ for $x \in A \cup B$; also $f(x)$ and $g(x)$ are both between $\varepsilon/2$ and $-\varepsilon/2$ for other x. So $|f(x) - g(x)| \leq \varepsilon$ for all x. Clearly $g(x) \neq 0$ for all x.

(b) There exist two disjoint closed sets A and B, not separated by the empty set. Define

$$f(x) = \frac{\operatorname{dist}(x, A) - \operatorname{dist}(x, B)}{\operatorname{dist}(x, A) + \operatorname{dist}(x, B)}.$$

Since the denominator never vanishes, this is a continuous function of x. Now suppose $g \colon S \to \mathbf{R}$ satisfies $\rho_u(f, g) \leq 1/2$, but $g(x) = 0$ has no solution. Then the set

$$U = \{\, x \in S : g(x) < 0 \,\} = \{\, x \in S : g(x) \leq 0 \,\}$$

is clopen, $U \supseteq A$, and $U \cap B = \varnothing$. This contradiction establishes the result. ☺

This result is the first case of the following, which is left to the reader.

(3.4.10) EXERCISE. *Let S be a metric space, and let $n \geq 0$ be an integer. The following are equivalent:*

  (1) *If $f_i \colon S \to \mathbf{R}$ are continuous functions for $i = 1, 2, \cdots, n$, then, for any $\varepsilon > 0$, there exist continuous functions $g_i \colon S \to \mathbf{R}$ with $\rho_u(f_i, g_i) \leq \varepsilon$ but the system $g_i(x) = 0$ ($i = 1, 2, \cdots, m$) has no solution.*

  (2) *If $A_1, B_1, A_2, B_2, \cdots A_n, B_n$ are $2n$ closed sets in S with $A_i \cap B_i = \varnothing$ for all i, then there exist sets $L_1, L_2, \cdots, L_n$ such that L_i separates A_i and B_i for all i, and $\bigcap_{i=1}^{n} L_i = \varnothing$.*

*3.5. REMARKS

The classical reference on topological dimension is the book by W. Hurewicz and H. Wallman [30]. This chapter is substantially based on their treatment. Topologists usually write "dim" for the covering dimension. But in this book, that is reserved for the Hausdorff dimension.

The small inductive dimension uses the catch-all symbol "∞" for spaces that do not have small inductive dimension otherwise specified by the definition. There is the possibility of a more refined classification of metric spaces of infinite dimension. The way in which the definition is formulated makes the use of transfinite ordinal numbers quite natural. If α is an ordinal number, and S is a metric space, then we say that ind $S \leq \alpha$ iff there is a base $\mathcal{B}$ for the open sets of S such that ind $\partial U < \alpha$ for $U \in \mathcal{B}$. Questions: Which ordinals α are the dimension of some metric space? Separable space? Compact space? Do we still need an extra symbol ∞, or does every metric space admit some ordinal as its dimension? What happens to the formula ind $(A \cup B) \leq$ ind $A +$ ind $B + 1$ (when addition is not commutative)? References: [18, p. 50], [30, p. 83].

The proof in Section 3.3 that ind $\mathbb{R}^2 = 2$ is a small taste of the mathematical field known as **algebraic topology**. A ten-cent description of algebraic topology might say that an algebraic object is associated with a situation from topology, in such a way that useful information can be obtained about the topology from the algebra. In the proof given in Section 3.3, the algebraic object that is used is the integers modulo 2. This is not a very sophisticated algebraic object, but it is enough to distinguish between two kinds of functions from the circle to itself.

Exercise 3.1.7: it is not true that the intersections of the gasket S with the small triangles that make up the approximations S_n constitute a base for the open sets. But it is almost true. (You also need unions of two such triangles.)

For Exercise 3.2.3: (1) $U = \{ x : \mathrm{dist}(x, A) < \mathrm{dist}(x, B) \}$.
(2) $U = \{ x \in S : \mathrm{dist}(x, A) < \mathrm{dist}(x, B)/2 \}$ and
$V = \{ x \in S : \mathrm{dist}(x, A) > 2\,\mathrm{dist}(x, B) \}$.

Hint for Exercise 3.4.10, (2) $\Longrightarrow$ (1). Given the functions $f_1, f_2, \cdots, f_n$, write $A_i = \{ x : f_i(x) \geq \varepsilon/2 \}$ and $B_i = \{ x : f_i(x) \leq -\varepsilon/2 \}$. By (2) there are disjoint open sets $U_i \supseteq A_i$ and $V_i \supseteq B_i$ such that the sets $L_i = S \setminus (U_i \cup V_i)$ satisfy $\bigcap_{i=1}^{n} L_i = \varnothing$. Then define $g_i \colon S \to \mathbb{R}$ so that $g_i(x) = f_i(x)$ on $A_i \cup B_i$, $0 < g_i(x) \leq \varepsilon/2$ on $U_i \setminus A_i$, and $0 > g_i(x) \geq -\varepsilon/2$ on $V_i \setminus B_i$.

A line is breadthless length.
A surface is that which has length and breadth only.
A solid is that which has length, breadth, and depth.
—Euclid, *The Elements* (translation by T. L. Heath)

---

*An optional section.

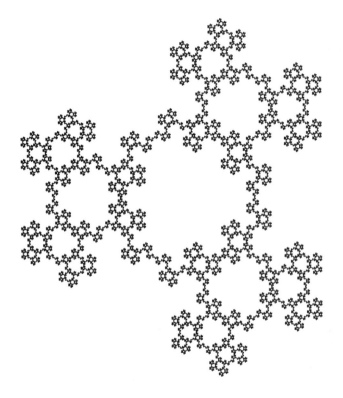

4

Self-Similarity

There are several variant notions of "dimension" that may be classified as fractal dimensions. The most widely used is known as the Hausdorff dimension. It will be considered in Chapter 6. We begin here with the **similarity dimension**, a fractal dimension that is easier to define (but not as useful).

At the same time, we will formally discuss **iterated function systems**. This is an efficient way of specifying many of the sets that we will be interested in. It has been publicized in recent years by Michael Barnsley; see [**3**].

4.1. RATIO LISTS

A **ratio list** is a finite list of positive numbers, $(r_1, r_2, \cdots, r_n)$. An **iterated function system** realizing a ratio list $(r_1, r_2, \cdots, r_n)$ in a metric space S is a list $(f_1, f_2, \cdots, f_n)$, where $f_i \colon S \to S$ is a similarity with ratio r_i. A nonempty compact set $K \subseteq S$ is an **invariant set** for the iterated function system $(f_1, f_2, \cdots, f_n)$ iff $K = f_1[K] \cup f_2[K] \cup \cdots \cup f_n[K]$.

For example, the triadic Cantor dust (page 6) is an invariant set for an iterated function system realizing the ratio list $(1/3, 1/3)$. Or the Sierpiński gasket (page 8) is an invariant set for an iterated function system realizing the ratio list $(1/2, 1/2, 1/2)$.

The **dimension** associated with a ratio list $(r_1, r_2, \cdots, r_n)$ is the positive number s such that $r_1^s + r_2^s + \cdots + r_n^s = 1$.

(4.1.1) THEOREM. *Let $(r_1, r_2, \cdots, r_n)$ be a ratio list. Suppose each $r_i < 1$. Then there is a unique nonnegative number s satisfying*

$$\sum_{i=1}^{n} r_i^s = 1.$$

The number s is 0 if and only if $n = 1$.

PROOF: Consider the function $\Phi \colon [0, \infty) \to [0, \infty)$ defined by

$$\Phi(s) = \sum_{i=1}^{n} r_i^s.$$

Then Φ is a continuous function, $\Phi(0) = n \geq 1$ and $\lim_{s \to \infty} \Phi(s) = 0 < 1$. Therefore, by the intermediate value theorem, there is at least one value s with $\Phi(s) = 1$. The derivative of Φ is

$$\sum_{i=1}^{n} r_i^s \log r_i.$$

This is < 0, so Φ is strictly decreasing. Therefore there is only one solution s to $\Phi(s) = 1$. If $n > 1$, then $\Phi(0) > 1$, so $s \neq 0$. ☺

A ratio list $(r_1, r_2, \cdots, r_n)$ is called **contracting** (or **hyperbolic**) iff $r_i < 1$ for all i.

The number s is called the **similarity dimension** of a (nonempty compact) set K iff there is a finite "decomposition" of K

$$K = \bigcup_{i=1}^{n} f_i[K]$$

where $(f_1, f_2, \cdots, f_n)$ is an iterated function system of similarities realizing a ratio list with dimension s.

It is, of course, conceivable that a given set admits two different decompositions, and therefore two different similarity dimensions. So perhaps it would make more sense to call s the similarity dimension of the iterated function system. We will see later that, under the right circumstances, the similarity dimension of a set K coincides with the Hausdorff dimension of K, which is uniquely determined by the set K.

Consider, for example, an interval $[a, b]$. It is the union of two smaller intervals, $[a, (a+b)/2]$ and $[(a+b)/2, b]$. Each of the parts is similar to the whole set $[a, b]$, with ratio $1/2$. The dimension of the ratio list $(1/2, 1/2)$ is the solution s of the equation

$$2 \left(\frac{1}{2} \right)^s = 1,$$

so $s = 1$. The similarity dimension is 1.

Note, however, that we can also write $[0, 1] = [0, 2/3] \cup [1/3, 1]$, corresponding to ratio list $(2/3, 2/3)$, yielding a dimension larger than 1. So in order for the similarity dimension to be a characteristic of the set, we will need to limit overlap in some way. This is discussed in Section 6.3.

What do we want to do for a rectangle $[a, b] \times [c, d]$ in $\mathbb{R}^2$? It is the union of four rectangles with sides half the size (Plate 9). The ratio list is $(1/2, 1/2, 1/2, 1/2)$ and the similarity dimension is 2.

How about a closed ball $\overline{B_r}(a)$ in $\mathbb{R}^2$ (a "disk")? Some geometry shows that a disk is not the union of finitely many disks of smaller radius. This illustrates the largest drawback of the similarity dimension: it is not defined for many sets, even very simple sets.

Now let us consider a more interesting example. The triadic Cantor dust is the invariant set for an iterated function system realizing ratio list $(1/3, 1/3)$ (page 6). So the similarity dimension is the solution s of the equation

$$2 \left(\frac{1}{3}\right)^{s} = 1,$$

so† $s = \log 2/ \log 3$, or approximately 0.6309. This agrees with the assertion (page 2) that the dimension should be less than 1.

Next take the Sierpiński gasket. The ratio list is $(1/2, 1/2, 1/2)$, so the similarity dimension is $\log 3/ \log 2 \approx 1.585$. It was asserted (page 8) that the dimension should be between 1 and 2.

(4.1.2) EXERCISE. *Calculate similarity dimensions for other sets: the Heighway dragon (page 20); the Koch curve (page 18); the McWorter pentigree (page 24); the twindragon (page 30); the Eisenstein fractions (page 32); 120-degree dragon (page 23).*

In these examples, we begin with a set, and then try to find a corresponding iterated function system. It is often useful to do things the other way around: begin with an iterated function system, and use it to obtain a set.

(4.1.3) THEOREM. *Let S be a complete metric space, let $(r_1, r_2, \cdots, r_n)$ be a contracting ratio list, and let $(f_1, f_2, \cdots, f_n)$ be an iterated function system of similarities in S that realizes the ratio list. Then there is a unique nonempty compact invariant set for the iterated function system.*

PROOF: Consider the metric space $\mathcal{K}(S)$ of nonempty compact subsets of S, with the Hausdorff metric D. Since S is complete, so is $\mathcal{K}(S)$, by Theorem 2.4.4. Define a function $F: \mathcal{K}(S) \to \mathcal{K}(S)$ as follows:

$$F(A) = \bigcup_{i=1}^{n} f_i[A].$$

The continuous image of a compact set is compact (Theorem 2.2.15), and the union of finitely many compact sets is compact (Exercise 2.2.14); thus if A is compact, then so is $F(A)$.

I claim that F is a contraction map. Let

$$r = \max\{r_1, r_2, \cdots, r_n\}.$$

Clearly $r < 1$. I will show that

$$D\big(F(A), F(B)\big) \leq rD(A, B).$$

---

†Since I am a mathematician, when I write "log" it refers to the natural logarithm. But in fact for the quotient of two logarithms, as we have here, it doesn't matter what the base is.

Let $q > D(A, B)$ be given. If x is any element of $F(A)$, then $x = f_i(x')$ for some i and some $x' \in A$. Since $q > D(A, B)$, there is a point $y' \in B$ with $\rho(x', y') < q$. But then the point $y = f_i(y') \in F(B)$ satisfies $\rho(x, y) = r_i\rho(x', y') < rq$. This is true for all $x \in F(A)$, so $F(A)$ is contained in the rq-neighborhood of $F(B)$. Similarly, $F(B)$ is contained in the rq-neighborhood of $F(A)$. Therefore $D\big(F(A), F(B)\big) \leq rq$. This is true for every $q > D(A, B)$, so we have $D\big(F(A), F(B)\big) \leq rD(A, B)$.

Therefore we have a contraction map F defined on a complete metric space $\mathcal{K}(S)$. By the contraction mapping theorem (2.1.36), F has a unique fixed point. A fixed point of F is exactly the same thing as an invariant set for $(f_1, f_2, \cdots, f_n)$. ☺

This proof, together with Corollary 2.1.37, provides a "construction" for the invariant set:

(4.1.4) COROLLARY. *In the notation of Theorem 4.1.3, if A_0 is any nonempty compact set in S, and if*

$$A_{k+1} = \bigcup_{i=1}^{n} f_i[A_k]$$

for $k \geq 0$, then the sequence (A_k) converges in the Hausdorff metric to the invariant set of the iterated function system.

The invariant set of a contracting iterated function system is sometimes called the **attractor** of the iterated function system.

(4.1.5) EXERCISE. *Let $(r_1, r_2, \cdots, r_n)$ be a contracting ratio list. Suppose an iterated function system $(f_1, f_2, \cdots, f_n)$ consists not of similarities, but only of maps $f_i \colon S \to S$ satisfying*

$$\rho(f_i(x), f_i(y)) \leq r_i\,\rho(x, y).$$

Show that if S is complete, then there is a unique invariant set.

We have been talking about "iterated function systems". What does this have to do with "iteration"? Let (f_1, f_2, f_3) be the iterated function system associated with the Sierpiński gasket (page 9). The iteration that we will be interested in involves starting with any point $a \in \mathbb{R}^2$, and then repeatedly applying these three functions, in any order.

(4.1.6) EXERCISE. *Let $k_1, k_2, k_3, \cdots$ be an infinite sequence in the set $\{1, 2, 3\}$. Let $a \in \mathbb{R}^2$ be a point. Let the sequence (x_n) be defined by*

$$x_0 = a; \qquad x_n = f_{k_n}(x_{n-1}) \quad \text{for } n \geq 1.$$

Then: (1) Every cluster point of the sequence (x_n) belongs to the Sierpiński gasket S; (2) Every point of the Sierpiński gasket is a cluster of such a sequence (x_n) for some choice of k_i; (3) There is a point a and choice sequence k_i so that S is exactly equal to the set of all cluster points of (x_n).

A more sophisticated version of (3) says that a "random" choice of (k_i) will (with probability one) have cluster set S. This fact has sometimes been used to produce a picture of the invariant set for an iterated function system on a computer.

(4.1.7) EXERCISE. *Let S be a complete metric space, let $(r_1, r_2, \cdots, r_n)$ be a contracting ratio list, and let $(f_1, f_2, \cdots, f_n)$ be a realization of the ratio list in S. State and prove the appropriate version of Exercise 4.1.6 for this situation.*

Recall the situation from Section 1.6. Let b be a complex number, and let $D = \{d_1, d_2, \cdots d_k\}$ be a finite set of complex numbers, including 0. We are interested in representing complex numbers (or real numbers) in the number system they define.

Write F for the set of "fractions"; that is, numbers of the form

$$\sum_{j=-\infty}^{-1} a_j b^j.$$

Analysis of the set F is a "self-similar" set problem:

(4.1.8) EXERCISE. *The set F is nonempty and compact, and is the invariant set for an iterated function system of similarities. What is the dimension of this iterated function system?*

There are various ways that such a "number system" can be generalized. Consider the following way to define a set in $\mathbb{R}^d$. Let $b \in \mathbb{R}$ with $|b| > 1$ as before, but let D be a finite subset of $\mathbb{R}^d$, including 0. Then let F be the set of all vectors

$$\sum_{j=-\infty}^{-1} b^j a_j,$$

where $a_j \in D$. Is there an iterated function system to describe F?

The **Menger sponge** is a set in $\mathbb{R}^3$. (See Figure 4.1.9.) Begin with a cube (with the inside) of side 1. Subdivide it into 27 smaller cubes by trisecting the edges. The trema to remove consists of the center cube and the 6 cubes in the centers of the faces. That means 20 cubes remain. (The boundaries of these 20 cubes must also remain, so that the set will be compact.) Continue in the same way with the small cubes.

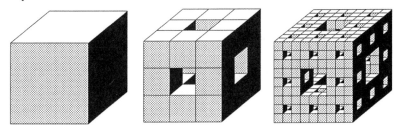

Figure 4.1.9. Menger sponge.

(4.1.10) EXERCISE. *What is the topological dimension of the Menger sponge? What is the similarity dimension of the Menger sponge?*

4.2. STRING MODELS

We will now consider "string models" in general. Given any ratio list, there is a realization that is, in an appropriate sense, the "best" realization (Theorem 4.2.3). Any other realization of the same ratio list is an "image" of this one. One advantage will be seen in Chapter 6 when Hausdorff dimension is computed.

Two instances of the string model have already been seen in Chapter 1, where the space of infinite strings in the alphabet $\{0, 1\}$ was used as a model for $[0, 1]$ and for the Cantor dust.

Example. Before we consider the general model, we will take one more example, the Sierpiński gasket. This time we will need to use an **infinite ternary tree**. Each node has exactly three children; each node except the root has exactly one parent, and every node is a descendent of the root. To represent this using strings, we will need a three-letter alphabet. Any three letters will do; I will use this alphabet: $E = \{\mathsf{L}, \mathsf{U}, \mathsf{R}\}$. (When I chose the letters I was looking at Figure 1.2.1, and thinking of the words "left", "upper", "right".)

We will write (as before) $E^{(k)}$ for the set of all k-letter strings from this alphabet; and $E^{(*)}$ for the set of all finite strings; and $E^{(\omega)}$ for the set of all infinite strings. We may identify $E^{(*)}$ with the infinite ternary tree: the empty string Λ is the root, and if α is a string, then $\alpha\mathsf{L}$ is the left child, $\alpha\mathsf{U}$ is the middle child (perhaps also called the "upper child"), $\alpha\mathsf{R}$ is the right child.

We want to define a map $h\colon E^{(\omega)} \to \mathbb{R}^2$ with range equal to the Sierpiński gasket S. The gasket itself is a union of three parts, the "left" part, the "upper" part, and the "right" part. There are three dilations of $\mathbb{R}^2$ corresponding to these three parts. They will be called now: f_L, f_U, f_R. The "model map" $h\colon E^{(\omega)} \to \mathbb{R}^2$ should be defined so that $h[E^{(\omega)}] = S$. It should be continuous and satisfy $h(\mathsf{L}\sigma) = f_\mathsf{L}(h(\sigma))$, $h(\mathsf{U}\sigma) = f_\mathsf{U}(h(\sigma))$, $h(\mathsf{R}\sigma) = f_\mathsf{R}(h(\sigma))$ for all strings σ.

We will describe this model map h using base 2 expansions in the (u, v) coordinate system of Exercise 1.2.4. A string from the alphabet $\{\mathsf{L}, \mathsf{U}, \mathsf{R}\}$ maps to a pair (u, v), according to the rules in the following table:

letter	digit of u	digit of v
L	0	0
U	0	1
R	1	0

For example,

$$h(\mathsf{LRLUU}\cdots) = \big((0.01000\cdots)_2, (0.00011\cdots)_2\big).$$

By Exercise 1.2.4, the range of h is exactly the Sierpiński gasket S.

The metric to be used on $E^{(\omega)}$ will be called $\rho_{1/2}$. If $\sigma, \tau \in E^{(\omega)}$, then

$$\rho_{1/2}(\sigma, \tau) = \left(\frac{1}{2}\right)^k$$

where k is the length of the longest common prefix of σ and τ. With this definition, we have $\mathrm{diam}\,[\alpha] = (1/2)^{|\alpha|}$. (Recall the notation $|\alpha|$ for the length of a finite string α.)

(4.2.1) EXERCISE. *The model map h defined above satisfies*

$$|h(\sigma) - h(\tau)| \le \rho_{1/2}(\sigma, \tau).$$

(4.2.2) EXERCISE. *Does the model map h have bounded decrease?*

General definition. Let $(r_1, r_2, \cdots, r_n)$ be a contracting ratio list. The model for this ratio list will be the space $E^{(\omega)}$ of infinite strings on an alphabet E with n letters. If no better choice suggests itself, the set $E = \{1, 2, \cdots, n\}$ may be used as the alphabet. Usually, however, we will choose letters that suggest the intended example. But there will be understood a one-to-one correspondence between the letters of E and the ratios in the ratio list. When the alphabet is known, we will often even use them to label the ratio list, so we may write $(r_e)_{e \in E}$ for $(r_1, r_2, \cdots, r_n)$.

For each letter $e \in E$, there is a corresponding function $\theta_e \colon E^{(\omega)} \to E^{(\omega)}$, called a **right shift**, defined by

$$\theta_e(\sigma) = e\sigma.$$

That is, insert the letter e at the beginning of the string. We will define a metric on $E^{(\omega)}$ so that the right shifts $(\theta_e)_{e \in E}$ form a realization of the given ratio list $(r_e)_{e \in E}$.

To define a metric on $E^{(\omega)}$, we will specify a diameter w_α for each node $\alpha \in E^{(*)}$ of the tree. This is done recursively:

$$w_\Lambda = 1,$$
$$w_{\alpha e} = w_\alpha r_e \qquad \text{for } \alpha \in E^{(*)} \text{ and } e \in E.$$

Alternatively, w_α is the product, over all the letters e that make up α, of the ratios r_e. In the Sierpiński gasket example above, for example, $w_{\mathsf{LRLU}} = r_{\mathsf{L}} r_{\mathsf{R}} r_{\mathsf{L}} r_{\mathsf{U}}$. The metric ρ is defined from these diameters in the usual way. If there are at least two letters, then there are no exceptions to the formula $\mathrm{diam}\,[\alpha] = w_\alpha$.

Now it is easy to verify that the iterated function system $(\theta_e)_{e \in E}$ realizes the ratio list $(r_e)_{e \in E}$: Suppose $\sigma, \tau \in E^{(\omega)}$ have longest common prefix α. If $e \in E$, then the longest common prefix of $e\sigma$ and $e\tau$ is $e\alpha$. So

$$\rho\big(\theta_e(\sigma), \theta_e(\tau)\big) = w_{e\alpha} = r_e w_\alpha = r_e\, \rho(\sigma, \tau).$$

That is, θ_e is a similarity on $\big(E^{(\omega)}, \rho\big)$ with ratio r_e.

The metric space $E^{(\omega)}$ is complete, so the right-shift realization $(\theta_e)_{e \in E}$ has a unique nonempty compact invariant set. In fact, that invariant set is the whole space $E^{(\omega)}$. The space $E^{(\omega)}$, together with the right shifts, will be called the **string model** of the ratio list $(r_e)_{e \in E}$.

(4.2.3) STRING MODEL THEOREM. *Let S be a nonempty complete metric space and let $(f_e)_{e \in E}$ be any iterated function system realizing the ratio list $(r_e)_{e \in E}$ in S. Assume that $r_e < 1$ for all e. Then there is a unique continuous function $h \colon E^{(\omega)} \to S$ such that*

$$h(e\sigma) = f_e(h(\sigma))$$

for all $\sigma \in E^{(\omega)}$ and $e \in E$. The range $h[E^{(\omega)}]$ is the invariant set of the iterated function system $(f_e)_{e \in E}$.

PROOF: We will use uniform convergence. We will define recursively a sequence (g_k) of continuous functions $g_k \colon E^{(\omega)} \to S$. Choose any point $a \in S$. Define $g_0(\sigma) = a$ for all σ. If g_k has been defined, then define g_{k+1} by:

$$g_{k+1}(e\sigma) = f_e(g_k(\sigma))$$

for $e \in E$ and $\sigma \in E^{(\omega)}$. The function g_0 is clearly continuous. We can verify by induction that g_{k+1} is continuous, using the fact that each of the sets $[e]$ is open.

I claim that the sequence (g_k) converges uniformly. Let $r = \max_e r_e$, so that $r < 1$. Now $E^{(\omega)}$ is compact, so $\rho_u(g_1, g_0)$ is finite. We have

$$\begin{aligned}
\rho(g_{k+1}(e\sigma), g_k(e\sigma)) &= \rho(f_e(g_k(\sigma)), f_e(g_{k-1}(\sigma))) \\
&\leq r_e \rho(g_k(\sigma), g_{k-1}(\sigma)) \\
&\leq r \rho_u(g_k, g_{k-1}).
\end{aligned}$$

Therefore $\rho_u(g_{k+1}, g_k) \leq r\rho_u(g_k, g_{k-1})$. So $\rho_u(g_{k+1}, g_k) \leq r^k \rho_u(g_1, g_0)$ by induction. By the triangle inequality, if $m \geq k$, then

$$\rho_u(g_m, g_k) \leq \sum_{j=k}^{m-1} \rho_u(g_{j+1}, g_j) \leq \sum_{j=k}^{\infty} r^j \rho_u(g_1, g_0).$$

This is the tail of a convergent geometric series, so it approaches 0 as $k \to \infty$. Therefore the sequence (g_k) is a Cauchy sequence in $\mathcal{C}(E^{(\omega)}, S)$. So it converges uniformly. Write h for its limit.

Now we have by the definition of the sequence (g_k),

$$g_{k+1}[E^{(\omega)}] = \bigcup_{e \in E} f_e[g_k[E^{(\omega)}]].$$

The sequence of sets $(g_k[E^{(\omega)}])$ converges to the invariant set by Corollary 4.1.4, and converges to $h[E^{(\omega)}]$ by Proposition 2.4.8. Therefore $h[E^{(\omega)}]$ is the invariant set.

For the uniqueness, suppose that $g \colon E^{(\omega)} \to S$ and $h \colon E^{(\omega)} \to S$ with $g(e\sigma) = f_e(g(\sigma))$ and $h(e\sigma) = f_e(h(\sigma))$. Now with $r = \max_e r_e$ as before, we have by the same calculation as above, $\rho_u(g, h) \leq r\rho_u(g, h)$. This is impossible unless $\rho_u(g, h) = 0$. So $g = h$. ☺

When the model map h is understood, we sometimes say that the infinite string $\sigma \in E^{(\omega)}$ is the **address** of the point $x = h(\sigma)$.

(4.2.4) EXERCISE. *The model map h has bounded increase.*

4.3. GRAPH SELF-SIMILARITY

There is a generalization of self-similarity that provides a way to study a larger class of sets. A definitive formulation is due to Mauldin and Williams, but variants were used by others. Figures 4.3.1 and 4.3.2 illustrate two examples. There will be a list of several nonempty compact sets to be constructed simultaneously. Each of them is decomposed into parts obtained from sets in the list using certain similarities.

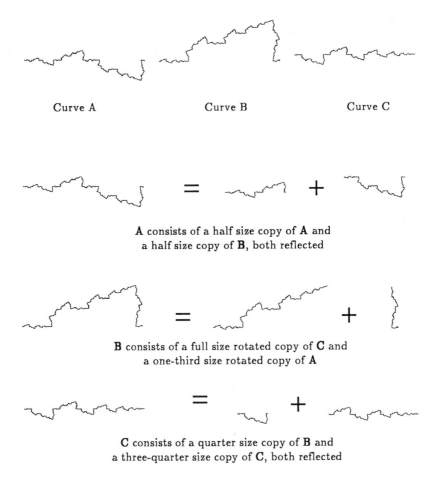

Curve A   Curve B   Curve C

A consists of a half size copy of **A** and
a half size copy of **B**, both reflected

B consists of a full size rotated copy of **C** and
a one-third size rotated copy of **A**

C consists of a quarter size copy of **B** and
a three-quarter size copy of **C**, both reflected

Figure 4.3.1. The MW curves.

An illustration (as in Figure 4.3.3 for the **golden rectangle fractal**) can be used to specify the information on how the similarities are to be made.

S   T

Figure 4.3.2. The golden rectangle fractal.

For each such construction, there corresponds in a natural way a directed multigraph, with a positive number associated with each edge. There is one node of the graph for each of the sets to be constructed. The edges from a node correspond to the subsets into which the corresponding set is decomposed, and the number associated with each edge corresponds to the ratio of the similarity. (Be careful to pay attention to which way the arrows go. It may seem more natural for some purposes to do it the other way around, but this is the direction chosen by Mauldin and Williams.) The structure involved, a directed multigraph (V, E, i, t) together with a function $r \colon V \to (0, \infty)$, will be called a **Mauldin-Williams graph**.

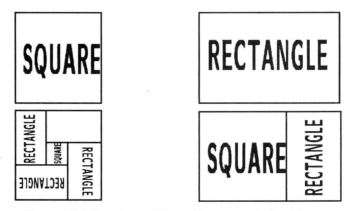

Figure 4.3.3. Structure of the golden rectangle fractal.

Suppose (V, E, i, t, r) is a Mauldin-Williams graph. An iterated function system realizing the graph is made up of metric spaces S_v, one for each vertex v, and similarities f_e, one for each edge $e \in E$, such that $f_e \colon S_v \to S_u$ if $e \in E_{uv}$, and f_e has ratio $r(e)$. An **invariant list** for such an iterated function system is

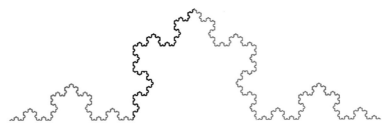

Plate 1

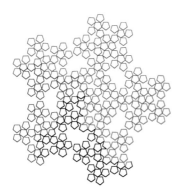
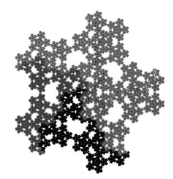

Plate 2

Plate 3

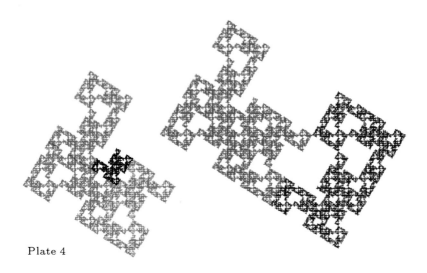

Plate 4

Plate 5

Plate 6

Plate 7

Plate 8

Plate 9

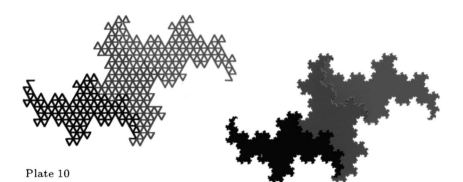

Plate 10

Plate 11

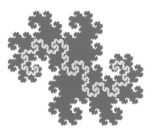

Plate 12

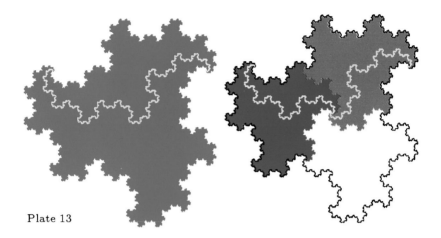

Plate 13

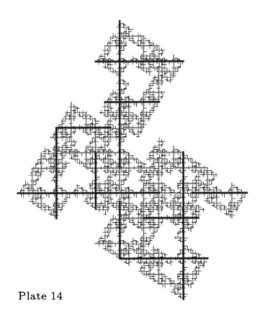

Plate 14

Plate 15

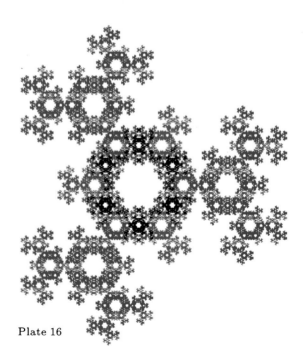

Plate 16

a list of nonempty compact sets $K_v \subseteq S_v$, one for each node $v \in V$, such that

$$K_u = \bigcup_{\substack{v \in V \\ e \in E_{uv}}} f_e[K_v]$$

for all $u \in V$.

Compare this abstract definition to the two examples. The Mauldin-Williams graphs are shown in Figures 4.3.4a and 4.3.4b.

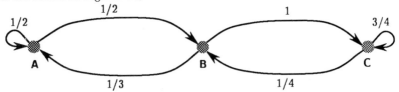

Figure 4.3.4a. Graph for the MW curves.

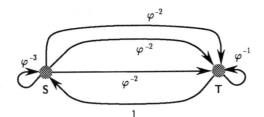

Figure 4.3.4b. Graph for the golden rectangle fractal.

Each of the nonempty compact sets K_v satisfying such equations will be said to have **graph self-similarity**.

For this purpose, the ratio lists discussed above correspond to graphs with a single vertex, and one loop for each item in the ratio list.

Existence of invariant lists. Just as in the case of ratio lists, if proper conditions are satisfied, then the invariant list of sets exists and is unique.

(4.3.5) THEOREM. Let (V, E, i, t, r) be a Mauldin-Williams graph. Suppose $r(e) < 1$ for all $e \in E$. Let $(f_e)_{e \in E}$ realize the graph in complete metric spaces S_v. Then there is a unique list $(K_v)_{v \in V}$ of nonempty compact sets $(K_v \subseteq S_v)$ such that

$$K_u = \bigcup_{\substack{v \in V \\ e \in E_{uv}}} f_e[K_v]$$

for all $u \in V$.

PROOF: The proof is similar to the proof of Theorem 4.1.3. The metric spaces $\mathcal{K}(S_v)$ are complete. So the finite Cartesian product

$$\prod_{v \in V} \mathcal{K}(S_v)$$

is also complete with the metric given by the maximum of the coordinate metrics. Let us write $(A_v)_{v \in V}$ for a typical element of this product space. The function defined by

$$F\big((A_v)_{v \in V}\big) = \left(\bigcup_{\substack{v \in V \\ e \in E_{uv}}} f_e[A_v] \right)_{u \in V}$$

is a contraction mapping, and its unique fixed point is the invariant list required. ☺

A Mauldin-Williams graph (V, E, i, t, r) will be called **strictly contracting** if the conditions $r(e) < 1$ are satisfied. The reader may have noticed that the definition of the golden rectangle fractal is not strictly contracting. See Exercise 4.3.8.

(4.3.6) EXERCISE. *Discuss the graph self-similarity of the boundary of Heighway's dragon.*

Path models. Let us consider the Mauldin-Williams graphs, and their realizations as iterated function systems of similarities. It will be useful to have a tree-type model for such a system. (Computing the Hausdorff dimension of the tree model will be much of the work in computing the Hausdorff dimension for sets with graph self-similarity.) The models will be analogous to the string models considered above. But (because of their construction) we will usually call them "path models".

Let (V, E, i, t, r) be a Mauldin-Williams graph. There is a "path forest" corresponding to it, as on page 74. The definition of the edge-value function r can be extended to paths by defining: $r(\Lambda_u) = 1$ for empty paths, and $r(\alpha e) = r(\alpha) r(e)$ for any path α and any edge e with $t(\alpha) = i(e)$.

Given an iterated function system $(f_e)_{e \in E}$, acting on spaces S_v, we may extend the notation to paths: f_{Λ_u} is the identity function on S_u, and $f_{\alpha e}$ is the composite function $f_\alpha \circ f_e$, defined by $f_{\alpha e}(x) = f_\alpha\big(f_e(x)\big)$. In cases when confusion will be minimal, we will save writing by using the edge e itself to stand for the function f_e; that is, we may write $e(x)$, where $x \in S_{t(e)}$, as an abbreviation of $f_e(x)$. Similarly, if α is a finite path, we may write $\alpha(x)$ for $f_\alpha(x)$.

Let (V, E, i, t, r) be a strictly contracting Mauldin-Williams graph. The spaces $E_v^{(\omega)}$ of infinite paths of the graph admit right-shift maps as before: If $e \in E_{uv}$, then we define $\theta_e : E_v^{(\omega)} \to E_u^{(\omega)}$ by $\theta_e(\sigma) = e\sigma$. One way to define metrics on the spaces $E_v^{(\omega)}$ so that this family of maps realizes the Mauldin-Williams graph is as follows: If $\sigma, \tau \in E_v^{(\omega)}$, and α is their longest common prefix, then $\rho(\sigma, \tau) = r(\alpha)$. [We will, however, use a different system of metrics in Section 6.4.]

(4.3.7) EXERCISE. *With the metrics ρ defined, the right-shift θ_e is a similarity with ratio $r(e)$.*

Rescaling. Suppose (V, E, i, t, r) is a Mauldin-Williams graph. Let maps f_e realize it in spaces (S_v, ρ_v). To **rescale** the realization, we replace each of the metrics ρ_v by a constant multiple $\rho'_v = a_v \rho_v$ of itself. The sets S_v are unchanged, and the maps f_e are unchanged. Of course an invariant list for the original iterated function system is exactly the same thing as an invariant list for the rescaled iterated function system.

Under the new metrics, what happens to the maps f_e? If $e \in E_{uv}$, then

$$\rho'_v\big(f_e(x), f_e(y)\big) = a_v \rho_v \big(f_e(x), f_e(y)\big)$$
$$= a_v r(e) \rho_u(x, y)$$
$$= \frac{a_v r(e)}{a_u} \rho'_u(x, y).$$

Thus, with the new metrics, the maps f_e realize a Mauldin-Williams graph (V, E, i, t, r'), where

$$r'(e) = \frac{a_v}{a_u} r(e) \qquad \text{for } e \in E_{uv}.$$

The Mauldin-Williams graph (V, E, i, t, r') is called a **rescaling** of the graph (V, E, i, t, r). A Mauldin-Williams graph (V, E, i, t, r) will be called **contracting** iff it is a rescaling of a strictly contracting graph.

Theorem 4.3.5 shows that a realization of a contracting graph in complete spaces has a unique invariant list. That list can be constructed by the method analogous to Corollary 4.1.4. This applies, for example to the golden rectangle fractal:

(4.3.8) EXERCISE. *The Mauldin-Williams graph of Figure 4.3.4b is contracting.*

Finally, we have an alleged criterion for contractivity.

(4.3.9) EXERCISE. *Prove or disprove: A Mauldin-Williams graph (V, E, i, t, r) is contracting if and only if $r(\alpha) < 1$ for all nonempty cycles $\alpha \in E^{(*)}$.*

Dimension. A Mauldin-Williams graph has a "dimension" associated with it, in the same way as a ratio list has a dimension associated with it. The case of a Mauldin-Williams graph with 2 nodes will be discussed now. The definitions for graphs with more than 2 nodes are given in Section 6.4.

Let (V, E, i, t, r) be a Mauldin-Williams graph. Suppose $V = \{1, 2\}$. Write

$$A(s) = \sum_{e \in E_{11}} r(e)^s$$
$$B(s) = \sum_{e \in E_{12}} r(e)^s$$
$$C(s) = \sum_{e \in E_{21}} r(e)^s$$
$$D(s) = \sum_{e \in E_{22}} r(e)^s,$$

and let

$$\Phi(s) = \frac{A(s) + D(s) + \sqrt{(A(s) - D(s))^2 + 4B(s)C(s)}}{2}.$$

The **dimension** of the graph is the solution s of the equation $\Phi(s) = 1$.

Of course (as in the case of a ratio list) such a number s need not exist. But with reasonable additional conditions, the dimension exists and is unique.

(4.3.10) PROPOSITION. *A strictly contracting, strongly connected Mauldin-Williams graph with 2 nodes has a unique dimension.*

PROOF: Since the graph is strictly contracting, we have $A(s) \to 0$, $B(s) \to 0$, $C(s) \to 0$, and $D(s) \to 0$ as $s \to \infty$, so $\Phi(s) \to 0$.

Since the graph is strongly connected, there is at least one edge from 1 to 2 and at least one edge from 2 to 1. So $B(0) \geq 1$ and $C(0) \geq 1$. Then $\Phi(0) \geq 1$. Equality $\Phi(0) = 1$ holds only if $A(0) = D(0) = 0$ and $B(0) = C(0) = 1$; that is, the two edges postulated are the only edges. In that case the dimension is 0. In all other cases, $\Phi(0) > 1$. Thus: if the graph is strictly contracting and strongly connected, then there is a nonnegative solution s to the equation $\Phi(s) = 1$.

I claim that the solution is unique since $\Phi'(s) < 0$ for all s. The partial derivatives of Φ are

$$\frac{\partial \Phi}{\partial A} = \frac{1}{2} + \frac{1}{2} \frac{A - D}{\sqrt{(A - D)^2 + 4BC}},$$

$$\frac{\partial \Phi}{\partial B} = \frac{C}{\sqrt{(A - D)^2 + 4BC}},$$

$$\frac{\partial \Phi}{\partial C} = \frac{B}{\sqrt{(A - D)^2 + 4BC}},$$

$$\frac{\partial \Phi}{\partial D} = \frac{1}{2} + \frac{1}{2} \frac{D - A}{\sqrt{(A - D)^2 + 4BC}}.$$

Recall that $A(s) \geq 0$, $B(s) > 0$, $C(s) > 0$, and $D(s) \geq 0$. Then $|A - D| \leq \sqrt{(A - D)^2 + 4BC}$, so

$$\frac{\partial \Phi}{\partial A} \geq \frac{1}{2} - \frac{1}{2} = 0,$$

$$\frac{\partial \Phi}{\partial B} > 0,$$

$$\frac{\partial \Phi}{\partial C} > 0,$$

$$\frac{\partial \Phi}{\partial D} \geq \frac{1}{2} - \frac{1}{2} = 0.$$

The four derivatives satisfy $A'(s) \leq 0$, $B'(s) < 0$, $C'(s) < 0$, and $D'(s) \leq 0$. So we have $\Phi'(s) < 0$. ☺

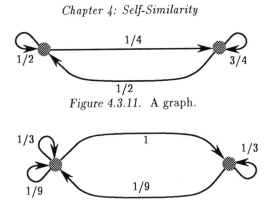

Figure 4.3.11. A graph.

Figure 4.3.12. A graph.

We will do examples later involving the Mauldin-Williams graphs in Figures 4.3.11 and 4.3.12.

(4.3.13) EXERCISE. *Compute the dimension of the graph of Figure 4.3.11.*

(4.3.14) EXERCISE. *Compute the dimension of the graph of Figure 4.3.12.*

Here is the way in which the dimension of a graph will be used.

(4.3.15) PROPOSITION. *Let (V, E, i, t, r) be a Mauldin-Williams graph with $V = \{1, 2\}$. The number $s \geq 0$ is the dimension of the graph if and only if there exist positive numbers x and y satisfying*

$$x = A(s)x + B(s)y$$
$$y = C(s)x + D(s)y.$$

PROOF: Write $A = A(s)$, etc. They satisfy $A \geq 0$, $B > 0$, $C > 0$, and $D \geq 0$. I must show that $(A + D + \sqrt{(A - D)^2 + 4BC})/2 = 1$ if and only if there exist positive x and y with $Ax + By = x$, $Cx + Dy = y$.

First, suppose $(A + D + \sqrt{(A - D)^2 + 4BC})/2 = 1$. Then $A + D < 2$, so either $A < 1$ or $D < 1$. We will take the case $A < 1$; the other case is similar. Let $x = B$ and $y = 1 - A$. Then $x > 0$ and $y > 0$. The first equation is $Ax + By = AB + B(1 - A) = B = x$. Algebra applied to the equation $(A + D + \sqrt{(A - D)^2 + 4BC})/2 = 1$ yields $BC - AD = 1 - A - D$. For the second equation, $Cx + Dy = CB + D(1 - A) = BC - AD + D = 1 - A = y$.

Conversely, suppose positive numbers x and y exist with $Ax + By = x$ and $Cx + Dy = y$. Since $y \neq 0$, we may solve:

$$\frac{B}{1 - A} = \frac{x}{y} = \frac{1 - D}{C}.$$

This means that $(1-A)(1-D) = BC$, so with a little algebra, $(A-D)^2 + 4BC = (2 - A - D)^2$. Now $B/(1 - A) = x/y > 0$, and $B > 0$, so $A < 1$. Similarly $D < 1$. So $2 - A - D > 0$, and therefore $\sqrt{(A - D)^2 + 4BC} = 2 - A - D$, or $(A + D + \sqrt{(A - D)^2 + 4BC})/2 = 1$. ☺

(4.3.16) EXERCISE. *How is the dimension of a two-node Mauldin-Williams graph affected by rescaling?*

(4.3.17) EXERCISE. *State and prove a "Path Model Theorem", analogous to Theorem 4.2.3, for graph self-similarity.*

*4.4. REMARKS

The idea of self-similarity is explored by P. A. P. Moran [42] and by John Hutchinson [31]. It will be very useful in Chapter 6 in computation of Hausdorff dimensions. Theorem 4.3.5 on the existence of the invariant set is taken from [31].

Graph self-similarity is a generalization of self-similarity. There are several sources for similar ideas; the one used as the model here is [39]. The examples shown in Figures 4.3.1 and 4.3.2 were concocted to illustrate the idea.

The Menger sponge is a universal 1-dimensional space. Karl Menger defined this space (and corresponding spaces in higher dimensions) for this purpose. The exact statement is like Theorem 3.1.9: Let S be a separable metric space. Then ind $S \leq 1$ if and only if S is homeomorphic to a subset of the Menger sponge. [7, p. 503*ff*]

Exercise 4.1.2: Approximate values of the similarity dimension: Heighway, 2; Koch, 1.26; McWorter, 1.86; twindragon, 2; Eisenstein, 2; 120-degree, 1.26.

Exercise 4.1.8: $\log k / \log |b|$.

Exercise 4.1.10: Topological dimension 1; similarity dimension approximately 2.73.

Exercise 4.3.13: 1.

Exercise 4.3.14: The dimension is $-\log x / \log 3$, where x is a solution of $x^3 - x^2 - 2x + 1 = 0$. So the dimension is approximately 0.737.

He had finally made some progress on the Barnsleyformer; because the trick wasn't in the morphogenesis after all, but in the fractal geometry.
—Michael F. Flynn, "Remembered Kisses", *ANALOG*, December, 1988.

---

*An optional section.

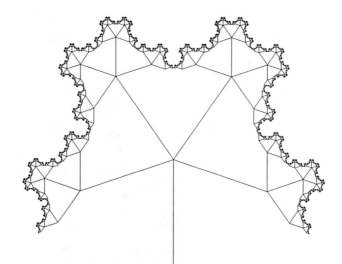

5

Measure Theory

This chapter contains the background from measure theory that is required to understand the Hausdorff dimension. It is true that the Hausdorff dimension can be defined in half a page without reference to measure theory, but when it is done that way there is no indication of the motivation for the definition.

Measure theory will also be indispensable in many of the proofs related to Hausdorff dimension. A proper use of measure theory will simplify many of the proofs of lower bounds. Instead of repeating a combinatorial calculation in each instance, we do the combinatorics once and for all in this chapter, and then repeatedly reap the benefits in Chapter 6.

Since measure theory (like metric topology) is a standard part of graduate mathematics curriculum today, most of the introductory remarks to Chapter 2 are also applicable here.

5.1. LEBESGUE MEASURE

Certain calculations will be done with the symbols ∞ and $-\infty$. They are not real numbers, but they can be useful in connection with calculations involving real numbers. Most of the conventions are sensible when you think about them. Here are some examples:

(1) If $a \in \mathbf{R}$, then $-\infty < a < \infty$.
(2) If $a \in \mathbf{R}$, then $a + \infty = \infty$ and $a - \infty = -\infty$. Also $\infty + \infty = \infty$ and $-\infty - \infty = -\infty$. The combination $\infty - \infty$ is not defined.
(3) If $a \in \mathbf{R}$ is positive, then $a \cdot \infty = \infty$ and $a \cdot (-\infty) = -\infty$. If $a \in \mathbf{R}$ is negative, then $a \cdot \infty = -\infty$ and $a \cdot (-\infty) = \infty$. The combination $0 \cdot \infty$ is not defined. [However, we do understand that an infinite series $\sum_{n=1}^{\infty} a_n$, where every term $a_n = 0$, has sum 0.]

The **length** of one of the intervals

$$(a, b) \quad (a, b] \quad [a, b) \quad [a, b]$$

is $b - a$, where $a, b \in \mathbf{R}$ and $a < b$. The length of the degenerate interval $[a, a] = \{a\}$ is 0; the length of the empty set $\varnothing$ is 0. The length of an unbounded interval

$$(a, \infty) \quad [a, \infty) \quad (-\infty, b) \quad (-\infty, b] \quad (-\infty, \infty)$$

is ∞. This follows the conventions on calculation with ∞.

We will be interested in a substantial generalization of the notion of the "length" of a subset of $\mathbf{R}$. The lemma that makes it possible asserts that the length of a countable union of intervals cannot exceed the sum of the lengths of the parts.

(5.1.1) LEMMA. *Suppose the closed interval* $[c, d]$ *is covered by a countable family of open intervals:*

$$[c, d] \subseteq \bigcup_{i \in \mathbf{N}} (a_i, b_i).$$

Then

$$d - c < \sum_{i=1}^{\infty} (b_i - a_i).$$

PROOF: First, since $[c, d]$ is a compact set, it is in fact covered by a finite number of the intervals:

$$[c, d] \subseteq \bigcup_{i=1}^{n} (a_i, b_i)$$

for some n. I will show that when this happens, the conclusion

$$d - c < \sum_{i=1}^{n} (b_i - a_i)$$

follows. The proof is by induction on n.

If $n = 1$, then $[c, d] \subseteq (a_1, b_1)$, so $a_1 < c$ and $d < b_1$. Thus $d - c < b_1 - a_1$, as required.

Now suppose $n \geq 2$, and the result is true for covers by at most $n - 1$ open intervals. Suppose

$$[c, d] \subseteq \bigcup_{i=1}^{n} (a_i, b_i).$$

If some interval (a_i, b_i) is disjoint from $[c, d]$, it may be omitted from the cover; then we have a cover by at most $n - 1$ sets, so we would be finished by the induction hypothesis. So assume $(a_i, b_i) \cap [c, d] \neq \varnothing$ for all i. Among all of the left endpoints a_i, there is one that is no larger than any of the others. By renumbering the intervals, let us assume that it is a_1. Since c is covered, we must have $a_1 < c$. Now if $b_1 > d$, we have $d - c < b_1 - a_1 \leq \sum_{i=1}^{n} (b_i - a_i)$, so we are finished. So suppose $b_1 \leq d$. Since (a_1, b_1) intersects $[c, d]$, we have $b_1 > c$. So $b_1 \in [c, d]$. At least one of the open intervals (a_i, b_i) covers the point b_1. By renumbering, we may assume it is (a_2, b_2). Finally, we have a cover of $[c, d]$ by $n - 1$ sets:

$$[c, d] \subseteq (a_1, b_2) \cup \bigcup_{i=3}^{n} (a_i, b_i).$$

So by the induction hypothesis,

$$d - c < (b_2 - a_1) + \sum_{i=3}^{n} (b_i - a_i)$$

$$\leq (b_2 - a_2) + (b_1 - a_1) + \sum_{i=3}^{n} (b_i - a_i)$$

as required. This completes the proof by induction. ☺

A useful generalization of the notion of the length of a subset of $\mathbb{R}$ is the **Lebesgue measure** of the set. This will be defined in stages. We will use half-open intervals of the form $[a, b)$ in the definition. Intervals of other forms could be used instead, but these have been chosen because of this convenient property:

(5.1.2) LEMMA. *Let $a < b$ be real numbers, and $\varepsilon > 0$. Then $[a, b)$ can be written as a finite disjoint union*

$$[a, b) = \bigcup_{i=1}^{n} [a_i, b_i),$$

with $b - a = \sum_{i=1}^{n} (b_i - a_i)$ and $b_i - a_i \leq \varepsilon$ for all i.

PROOF: Choose $n \in \mathbb{N}$ so large that $(b - a)/n \leq \varepsilon$. Let $b_i = a + i(b - a)/n$ for $0 \leq i \leq n$, and $a_i = b_{i-1}$. ☺

Now let A be any subset of $\mathbb{R}$. The **Lebesgue outer measure** of A is obtained by covering A with countably many half-open intervals of total length as small as possible. In symbols,†

$$\overline{\mathcal{L}}(A) = \inf \sum_{j=1}^{\infty} (b_j - a_j)$$

where the infimum is over all countable families $\{ [a_j, b_j) : j \in \mathbb{N} \}$ of half-open intervals with $A \subseteq \bigcup_{j \in \mathbb{N}} [a_j, b_j)$.

---

†In case you can't tell, the symbol $\mathcal{L}$ is supposed to be a fancy letter L, for "Lebesgue".

(5.1.3) LEMMA. *Let $A \subseteq \mathbf{R}$ and let $\varepsilon > 0$. Then*

$$\overline{\mathcal{L}}(A) = \inf \sum_{j=1}^{\infty} (b_j - a_j)$$

where the infimum is over all countable families $\{[a_j, b_j) : j \in \mathbf{N}\}$ of half-open intervals with $A \subseteq \bigcup_{j \in \mathbf{N}}[a_j, b_j)$ and $b_j - a_j \leq \varepsilon$ for all j.

PROOF: This follows from Lemma 1.5.2. ☺

We must do some combinatorics on the line to see that the definition is not trivial.*

(5.1.4) THEOREM. *If A is an interval, then $\overline{\mathcal{L}}(A)$ is the length of A.*

PROOF: Suppose $A = [a, b]$, where $a < b$ are real numbers. First, if $\varepsilon > 0$, then the singleton $\{[a, b + \varepsilon)\}$ covers the set A, so $\overline{\mathcal{L}}(A) \leq b - a + \varepsilon$. This is true for any $\varepsilon > 0$, so $\overline{\mathcal{L}}(A) \leq b - a$.

Now suppose $A \subseteq \bigcup_{j \in \mathbf{N}}[a_j, b_j)$. Let $\varepsilon > 0$, and write $a_j' = a_j - \varepsilon/2^j$. Then $A \subseteq \bigcup_{j \in \mathbf{N}}(a_j', b_j)$. By Lemma 5.1.1 $\sum_{j=1}^{\infty}(b_j - a_j') > b - a$. So we have $\sum_{j=1}^{\infty}(b_j - a_j) \geq \sum_{j=1}^{\infty}(b_j - a_j') - \varepsilon > b - a - \varepsilon$. This is true for any $\varepsilon > 0$, so $\sum_{j=1}^{\infty}(b_j - a_j) \geq b - a$. Therefore $\overline{\mathcal{L}}(A) \geq b - a$. So we have $\overline{\mathcal{L}}([a, b]) = b - a$.

Next consider $A = (a, b)$. Then $\overline{\mathcal{L}}(A) \leq \overline{\mathcal{L}}([a, b]) = b - a$ and on the other hand $\overline{\mathcal{L}}(A) \geq \overline{\mathcal{L}}([a + \varepsilon, b - \varepsilon]) = b - a - 2\varepsilon$ for any $\varepsilon > 0$. Similar arguments cover cases $[a, b)$ and $(a, b]$. If $A = [a, \infty)$, then $A \supseteq [a, a + t]$ for any $t > 0$, and therefore $\overline{\mathcal{L}}(A) \geq t$; this means that $\overline{\mathcal{L}}(A) = \infty$. Similar arguments cover the other cases of infinite length intervals. ☺

Here are some of the basic properties of Lebesgue outer measure.

(5.1.5) THEOREM.
   (1) $\overline{\mathcal{L}}(\varnothing) = 0$;
   (2) *if $A \subseteq B$, then $\overline{\mathcal{L}}(A) \leq \overline{\mathcal{L}}(B)$;*
   (3) $\overline{\mathcal{L}}\left(\bigcup_{n \in \mathbf{N}} A_n\right) \leq \sum_{n=1}^{\infty} \overline{\mathcal{L}}(A_n)$.

PROOF: For (1), note that $\varnothing \subseteq \bigcup_{i \in \mathbf{N}}[0, \varepsilon/2^i)$, so $\overline{\mathcal{L}}(\varnothing) \leq \varepsilon$. For (2), note that any cover of B is also a cover of A.

Now consider (3). If $\overline{\mathcal{L}}(A_n) = \infty$ for some n, then the inequality is clear. So suppose $\overline{\mathcal{L}}(A_n) < \infty$ for all n. Let $\varepsilon > 0$. For each n, choose a countable cover $\mathcal{D}_n$ of A_n by half-open intervals with

$$\sum_{D \in \mathcal{D}_n} \overline{\mathcal{L}}(D) \leq \overline{\mathcal{L}}(A_n) + 2^{-n}\varepsilon.$$

---

*I can easily write down the same definition for subsets of the rational numbers. But then every set turns out to have outer measure 0, so it is not a very useful definition.

Now $\mathcal{D} = \bigcup_{n \in \mathbb{N}} \mathcal{D}_n$ is a countable cover of the union $\bigcup_{n \in \mathbb{N}} A_n$. Therefore

$$\overline{\mathcal{L}}\left(\bigcup_{n \in \mathbb{N}} A_n\right) \le \sum_{D \in \mathcal{D}} \overline{\mathcal{L}}(D) \le \sum_{n=1}^{\infty} \sum_{D \in \mathcal{D}_n} \overline{\mathcal{L}}(D)$$

$$\le \sum_{n=1}^{\infty} \overline{\mathcal{L}}(A_n) + \sum_{n=1}^{\infty} 2^{-n} \varepsilon = \sum_{n=1}^{\infty} \overline{\mathcal{L}}(A_n) + \varepsilon.$$

Since ε was any positive number, we have

$$\overline{\mathcal{L}}\left(\bigcup_{n \in \mathbb{N}} A_n\right) \le \sum_{n=1}^{\infty} \overline{\mathcal{L}}(A_n). \ \ \odot$$

In general, the inequality in part (3) is not equality, even for two disjoint sets. But we do have equality in some cases. The simplest case is the following:

(5.1.6) THEOREM. *Let $A, B \subseteq \mathbb{R}$ with $\mathrm{dist}(A, B) > 0$. Then $\overline{\mathcal{L}}(A \cup B) = \overline{\mathcal{L}}(A) + \overline{\mathcal{L}}(B)$.*

PROOF: First, the inequality $\overline{\mathcal{L}}(A \cup B) \le \overline{\mathcal{L}}(A) + \overline{\mathcal{L}}(B)$ follows from part (3) of Theorem 5.1.5. Let $\varepsilon = \mathrm{dist}(A, B)/2$, and let $A \subseteq \bigcup_{j \in \mathbb{N}} [a_j, b_j)$, where $b_j - a_j \le \varepsilon$ for all j. Then each interval $[a_j, b_j)$ intersects at most one of the sets A and B. So the collection $\mathcal{D} = \{ [a_j, b_j) : j \in \mathbb{N} \}$ can be written as the disjoint union of two collections, $\mathcal{D} = \mathcal{D}_1 \cup \mathcal{D}_2$, where $\mathcal{D}_1$ covers A and $\mathcal{D}_2$ covers B. Now $\overline{\mathcal{L}}(A) \le \sum_{D \in \mathcal{D}_1} \overline{\mathcal{L}}(D)$ and $\overline{\mathcal{L}}(B) \le \sum_{D \in \mathcal{D}_2} \overline{\mathcal{L}}(D)$, so

$$\overline{\mathcal{L}}(A) + \overline{\mathcal{L}}(B) \le \sum_{D \in \mathcal{D}_1} \overline{\mathcal{L}}(D) + \sum_{D \in \mathcal{D}_2} \overline{\mathcal{L}}(D) = \sum_{D \in \mathcal{D}} \overline{\mathcal{L}}(D) \le \sum_{j=1}^{\infty} b_j - a_j.$$

Therefore, by Lemma 5.1.3, we have $\overline{\mathcal{L}}(A) + \overline{\mathcal{L}}(B) \le \overline{\mathcal{L}}(A \cup B)$. $\odot$

(5.1.7) COROLLARY. *If $A, B \subseteq \mathbb{R}$ are disjoint and compact, then we have $\overline{\mathcal{L}}(A) + \overline{\mathcal{L}}(B) = \overline{\mathcal{L}}(A \cup B)$.*

PROOF: Apply Theorems 2.2.19 and 5.1.6. $\odot$

(5.1.8) THEOREM. *If $A \subseteq \mathbb{R}$, then*

$$\overline{\mathcal{L}}(A) = \inf \left\{ \overline{\mathcal{L}}(U) : U \supseteq A, \ U \ open \right\}.$$

PROOF: Certainly $\overline{\mathcal{L}}(A) \le \inf \{ \overline{\mathcal{L}}(U) : U \supseteq A, \ U \text{ open} \}$. So I must prove the opposite inequality. If $\overline{\mathcal{L}}(A) = \infty$, it is trivially true. So suppose $\overline{\mathcal{L}}(A) < \infty$. Let $\varepsilon > 0$. Then there exists a cover $\bigcup_{j \in \mathbb{N}} [a_j, b_j)$ of A with $\sum_{j=1}^{\infty} (b_j - a_j) \le \overline{\mathcal{L}}(A) + \varepsilon/2$. Now the set $U = \bigcup_{j \in \mathbb{N}} (a_j - \varepsilon/2^{j+1}, b_j)$ is open, $U \supseteq A$, and $\overline{\mathcal{L}}(U) \le \sum_{j=1}^{\infty} (b_j - a_j) + \varepsilon/2 \le \overline{\mathcal{L}}(A) + \varepsilon$. Therefore $\overline{\mathcal{L}}(A) + \varepsilon \ge \overline{\mathcal{L}}(U)$. This shows that $\overline{\mathcal{L}}(A) \ge \inf \{ \overline{\mathcal{L}}(U) : U \supseteq A \}$. $\odot$

The outer measure $\overline{\mathcal{L}}(A)$ of a set $A \subseteq \mathbf{R}$ is determined by approximating a set from the outside by open sets. There is a corresponding "inner measure", obtained by approximating a set from the inside. This time, however, we will use compact sets.

Let $A \subseteq \mathbf{R}$. The **Lebesgue inner measure** of the set A is

$$\underline{\mathcal{L}}(A) = \sup \left\{ \overline{\mathcal{L}}(K) : K \subseteq A, K \text{ compact} \right\}.$$

Again, we need an argument to see that the definition is interesting.

(5.1.9) THEOREM. *If A is an interval, then $\underline{\mathcal{L}}(A)$ is the length of A.*

PROOF: We consider the case of an open interval $A = (a, b)$. Other kinds of intervals follow from this case as before.

If $K \subseteq A$ is compact, then K is covered by the single interval A, so that $\overline{\mathcal{L}}(K) \leq b - a$. Therefore $\underline{\mathcal{L}}(A) \leq b - a$. On the other hand, if $\varepsilon > 0$, then the set $[a + \varepsilon, b - \varepsilon]$ is compact, so $\underline{\mathcal{L}}(A) \geq \overline{\mathcal{L}}([a + \varepsilon, b - \varepsilon]) = b - a - 2\varepsilon$. This is true for any $\varepsilon > 0$, so $\underline{\mathcal{L}}(A) \geq b - a$. ☺

(5.1.10) EXERCISE. *If $A \subseteq \mathbf{R}$ is any set, then $\underline{\mathcal{L}}(A) \leq \overline{\mathcal{L}}(A)$.*

It is not possible to prove that $\underline{\mathcal{L}}(A) = \overline{\mathcal{L}}(A)$ in general. A set A is called **Lebesgue measurable**, roughly speaking, when this equation is true. Precisely: If $\overline{\mathcal{L}}(A) < \infty$, then A is Lebesgue measurable iff $\underline{\mathcal{L}}(A) = \overline{\mathcal{L}}(A)$. If $\overline{\mathcal{L}}(A) = \infty$, then A is Lebesgue measurable iff $A \cap [-n, n]$ is Lebesgue measurable for all $n \in \mathbf{N}$. If A is Lebesgue measurable, we will write $\mathcal{L}(A)$ for the common value of $\overline{\mathcal{L}}(A)$ and $\underline{\mathcal{L}}(A)$, and call it simply the **Lebesgue measure** of A.

(5.1.11) THEOREM. *Let $A_1, A_2, \cdots$ be disjoint Lebesgue measurable sets. Then $\bigcup_n A_n$ is measurable, and $\mathcal{L}(\bigcup_n A_n) = \sum_n \mathcal{L}(A_n)$.*

PROOF: It is enough to prove the theorem in the case that $\mathcal{L}(\bigcup A_n) < \infty$, since the general case will then follow by applying this case to sets $A_n \cap [-m, m]$. We know by Theorem 5.1.5 that $\overline{\mathcal{L}}(\bigcup A_n) \leq \sum \mathcal{L}(A_n)$. Let $\varepsilon > 0$. For each n, choose a compact set $K_n \subseteq A_n$ with $\overline{\mathcal{L}}(K_n) \geq \underline{\mathcal{L}}(A_n) - \varepsilon/2^n$. Since A_n is measurable, $\overline{\mathcal{L}}(K_n) \geq \overline{\mathcal{L}}(A_n) - \varepsilon/2^n$. Now the sets K_n are disjoint, so by Corollary 5.1.7, the compact set $L_m = K_1 \cup K_2 \cup \cdots \cup K_m$ satisfies $\overline{\mathcal{L}}(L_m) = \overline{\mathcal{L}}(K_1) + \cdots + \overline{\mathcal{L}}(K_m)$. Therefore $\underline{\mathcal{L}}(\bigcup A_n) \geq \sum_{n=1}^{m} \overline{\mathcal{L}}(K_n)$. Now this is true for all m, so $\underline{\mathcal{L}}(\bigcup A_n) \geq \sum_{n=1}^{\infty} \mathcal{L}(K_n) \geq \sum_{n=1}^{\infty} \mathcal{L}(A_n) - \varepsilon$. This is true for any positive ε, so we have $\underline{\mathcal{L}}(\bigcup_n A_n) \geq \sum \mathcal{L}(A_n)$.

So $\overline{\mathcal{L}}(\bigcup A_n) = \underline{\mathcal{L}}(\bigcup A_n)$, and therefore $\bigcup A_n$ is measurable and $\mathcal{L}(\bigcup A_n) = \sum \mathcal{L}(A_n)$. ☺

(5.1.12) THEOREM. *A compact set $K \subseteq \mathbf{R}$ is Lebesgue measurable. An open set $U \subseteq \mathbf{R}$ is Lebesgue measurable.*

PROOF: The compact set K is bounded, so $K \subseteq [-n, n]$ for some n, and therefore $\overline{\mathcal{L}}(K) < \infty$. The compact set K is a subset of K, so $\underline{\mathcal{L}}(K) \geq \overline{\mathcal{L}}(K)$.

Let U be an open set. It is enough to do the case $\overline{\mathcal{L}}(U) < \infty$. For each $x \in U$, there is an open interval I with $x \in I \subseteq U$. By the Lindelöf property, U is the union of countably many of these intervals, say $U = \bigcup_{j \in \mathbf{N}} I_j$. Now each set $I_n \setminus \bigcup_{j=1}^{n-1} I_j$ is a finite union of intervals (open, closed, half-open) so that U is a disjoint union of countably many intervals. So U is measurable. ☺

(5.1.13) THEOREM. *Let $A \subseteq \mathbf{R}$. Then A is measurable if and only if, for every $\varepsilon > 0$, there exist an open set U and a closed set F with $U \supseteq A \supseteq F$ and $\mathcal{L}(U \setminus F) < \varepsilon$.*

PROOF: Suppose first that A is measurable. We consider first of all the case $\mathcal{L}(A) < \infty$. Then there exists an open set $U \supseteq A$ such that $\mathcal{L}(U) < \mathcal{L}(A) + \varepsilon/2$. There exists a compact (therefore closed) set $F \subseteq A$ with $\mathcal{L}(F) > \mathcal{L}(A) - \varepsilon/2$. Now $U \setminus F$ is open, hence measurable, and F is compact, hence measurable, so $\mathcal{L}(U) = \mathcal{L}(U \setminus F) + \mathcal{L}(F)$. Since the terms are all finite, we may subtract, and we get

$$\mathcal{L}(U \setminus F) = \mathcal{L}(U) - \mathcal{L}(F) < \mathcal{L}(A) + \varepsilon/2 - \mathcal{L}(A) + \varepsilon/2 = \varepsilon.$$

Now we take the case $\mathcal{L}(A) = \infty$. All of the sets $A \cap [-n, n]$ are measurable. So there exist open sets $U_n \supseteq A \cap [-n, n]$ and compact sets $F_n \subseteq A \cap [-n, n]$ with $\mathcal{L}(U_n \setminus F_n) < \varepsilon/2^n$. Now $U = \bigcup U_n$ is open, and

$$F = \bigcup_{n \in \mathbf{N}} \left(F_n \cap ([-n, -n+1] \cup [n-1, n]) \right)$$

is closed (Exercise 2.1.42). We have $U \supseteq A \supseteq F$, and $U \setminus F \subseteq \bigcup_{n \in \mathbf{N}}(U_n \setminus F_n)$, so that $\mathcal{L}(U \setminus F) \leq \sum \mathcal{L}(U_n \setminus F_n) < \varepsilon$.

Conversely, suppose sets U and F exist. First assume $\overline{\mathcal{L}}(A) < \infty$. Then $\mathcal{L}(F) < \infty$, and $\mathcal{L}(U) \leq \mathcal{L}(U \setminus F) + \mathcal{L}(F) < \varepsilon + \mathcal{L}(F) < \infty$. Now $\overline{\mathcal{L}}(A) \leq \overline{\mathcal{L}}(U) < \mathcal{L}(F) + \varepsilon \leq \underline{\mathcal{L}}(A) + \varepsilon$. This is true for any $\varepsilon > 0$, so $\overline{\mathcal{L}}(A) = \underline{\mathcal{L}}(A)$, so A is measurable.

For the case $\overline{\mathcal{L}}(A) = \infty$, we have $U \cap (-n-\varepsilon, n+\varepsilon) \supseteq A \cap [-n, n] \supseteq F \cap [-n, n]$, and the previous case may be applied to these sets, using 3ε in place of ε. ☺

Here are the basic algebraic properties of Lebesgue measurable sets.

(5.1.14) THEOREM.

(1) *Both $\varnothing$ and $\mathbf{R}$ are Lebesgue measurable.*

(2) *If $A \subseteq \mathbf{R}$ is Lebesgue measurable, then so is its complement $\mathbf{R} \setminus A$.*

(3) *If A and B are measurable, then so are $A \cap B$, $A \cup B$, and $A \setminus B$.*

(4) *If A_n is measurable for $n \in \mathbf{N}$, then so are $\bigcup_{n \in \mathbf{N}} A_n$ and $\bigcap_{n \in \mathbf{N}} A_n$.*

PROOF: For (1), note that $\overline{\mathcal{L}}(\varnothing) = 0$ and $\mathbf{R} \cap [-n, n]$ is measurable for all n.

For (2), note that if $F \subseteq A \subseteq U$, then $\mathbf{R} \backslash U \subseteq \mathbf{R} \backslash A \subseteq \mathbf{R} \backslash F$ and $(\mathbf{R} \backslash F) \backslash (\mathbf{R} \backslash U) = U \backslash F$.

For the intersection in (3), note that if $F_1 \subseteq A \subseteq U_1$ and $F_2 \subseteq B \subseteq U_2$, then $F_1 \cap F_2 \subseteq A \cap B \subseteq U_1 \cap U_2$ and $(U_1 \cap U_2) \backslash (F_1 \cap F_2) \subseteq (U_1 \backslash F_1) \cup (U_2 \backslash F_2)$. This is enough to show that $A \cap B$ is measurable. Now $A \cup B = \mathbf{R} \backslash ((\mathbf{R} \backslash A) \cap (\mathbf{R} \backslash B))$, so $A \cup B$ is measurable. And $A \backslash B = A \cap (\mathbf{R} \backslash B)$, so $A \backslash B$ is measurable.

Finally, for (4), note that by (3) we may find disjoint measurable sets B_n with the same union as A_n, so that Theorem 5.1.11 is applicable. The intersection follows by taking complements. ☺

Note that (4) involves only *countable* unions and intersections.

(5.1.15) PROPOSITION. *The Lebesgue measure of the triadic Cantor dust is* 0.

PROOF: The Cantor dust C is constructed on page 2. The set $C_n \supseteq C$ consists of 2^n disjoint intervals of length 3^{-n}. Therefore $\mathcal{L}(C) \leq 2^n \cdot 3^{-n}$. This has limit 0, so $\mathcal{L}(C) = 0$. ☺

Carathéodory measurability. Carathéodory provided an alternate definition of measurability. Its disadvantage is that the motivation is not as clear. Its advantage is (as we will see in Section 5.2) that it can be used in other situations.

A set $A \subseteq \mathbf{R}$ is **Carathéodory measurable** iff $\overline{\mathcal{L}}(E) = \overline{\mathcal{L}}(E \cap A) + \overline{\mathcal{L}}(E \backslash A)$ for all sets $E \subseteq \mathbf{R}$.

(5.1.16) PROPOSITION. *A set* $A \subseteq \mathbf{R}$ *is Carathéodory measurable if and only if it is Lebesgue measurable.*

PROOF: Suppose A is Lebesgue measurable. Let E be a test set. The inequality $\overline{\mathcal{L}}(E) \leq \overline{\mathcal{L}}(E \cap A) + \overline{\mathcal{L}}(E \backslash A)$ is always true. Let $\varepsilon > 0$. There exist an open set U and a closed set F with $F \subseteq A \subseteq U$ and $\mathcal{L}(U \backslash F) < \varepsilon$. Let $V \supseteq E$ be an open set. Then

$$\begin{aligned}
\overline{\mathcal{L}}(E \backslash A) + \overline{\mathcal{L}}(E \cap A) &\leq \mathcal{L}(V \backslash F) + \mathcal{L}(V \cap U) \\
&\leq \mathcal{L}(V \backslash U) + \mathcal{L}(U \backslash F) + \mathcal{L}(V \cap U) \\
&< \mathcal{L}(V) + \varepsilon.
\end{aligned}$$

Now take the infimum over all such V, to get $\overline{\mathcal{L}}(E \cap A) + \overline{\mathcal{L}}(E \backslash A) < \overline{\mathcal{L}}(E) + \varepsilon$. Therefore $\overline{\mathcal{L}}(E \cap A) + \overline{\mathcal{L}}(E \backslash A) \leq \overline{\mathcal{L}}(E)$. This proves that A is Carathéodory measurable.

Conversely, suppose A is Carathéodory measurable. Consider the case in which $\overline{\mathcal{L}}(A) < \infty$. Let $\varepsilon > 0$. Let $U \supseteq A$ satisfy $\mathcal{L}(U) < \overline{\mathcal{L}}(A) + \varepsilon$. Now we have

$$\overline{\mathcal{L}}(U) = \overline{\mathcal{L}}(U \cap A) + \overline{\mathcal{L}}(U \backslash A),$$

so that $\overline{\mathcal{L}}(U \backslash A) < \varepsilon$. So there is an open set $V \supseteq U \backslash A$ with $\mathcal{L}(V) < \varepsilon$. Then $U \backslash V$ is Lebesgue measurable, and $\mathcal{L}(U \backslash V) > \mathcal{L}(U) - \varepsilon$, so there is a closed set $F \subseteq U \backslash V \subseteq A$ with $\mathcal{L}(F) > \mathcal{L}(U) - \varepsilon$. Thus $F \subseteq A \subseteq U$ and $\mathcal{L}(U \backslash F) < \varepsilon$. Therefore A is Lebesgue measurable. ☺

(5.1.17) THEOREM. *Let $A \subseteq \mathbf{R}$ be Lebesgue measurable, and let a similarity $f : \mathbf{R} \to \mathbf{R}$ with ratio r be given. Then $f[A]$ is Lebesgue measurable and $\mathcal{L}(f[A]) = r\mathcal{L}(A)$.*

PROOF: If $r = 0$, then the range of f is a single point, so of course $f[A]$ is measurable and $\mathcal{L}(f[A]) = 0$. So suppose $r > 0$.

Consider an interval $I = [a, b)$. The image is an interval, either $[f(a), f(b))$ or $(f(b), f(a)]$, with length $|f(b) - f(a)| = r|b-a|$. Therefore $\overline{\mathcal{L}}(f[I]) = r|b-a|$. Now if $A \subseteq \bigcup_{j \in \mathbf{N}} [a_j, b_j)$, then $f[A] \subseteq \bigcup f[[a_j, b_j)]$, so $\overline{\mathcal{L}}(f[A]) \leq \sum \overline{\mathcal{L}}(f[[a_j, b_j)]) = r \sum (b_j - a_j)$. Therefore we have $\overline{\mathcal{L}}(f[A]) \leq r\overline{\mathcal{L}}(A)$. If we apply the same thing to the inverse map f^{-1}, which is a similarity with ratio $1/r$, we get $\overline{\mathcal{L}}(f[A]) \geq r\overline{\mathcal{L}}(A)$. Therefore $\overline{\mathcal{L}}(f[A]) = r\overline{\mathcal{L}}(A)$

Now f is a homeomorphism, so the image of an open set is open and the image of a closed set is closed. If $A \subseteq \mathbf{R}$ is measurable, then, for every $\varepsilon > 0$, there exist a closed set F and an open set U with $F \subseteq A \subseteq U$ and $\overline{\mathcal{L}}(U \setminus F) < \varepsilon$. So we have $f[F] \subseteq f[A] \subseteq f[U]$ and $\overline{\mathcal{L}}(f[U] \setminus f[F]) < r\varepsilon$. So $f[A]$ is also measurable. ☺

Next is a preview of how measure theory is related to fractal dimension. In general, we do not yet know that the similarity dimension of a set is unique. However, we can now establish that in one situation we can determine the similarity dimension.

(5.1.18) EXERCISE. *Let $(r_1, r_2, \cdots, r_n)$ be a contracting ratio list; let s be its dimension; let $(f_1, f_2, \cdots f_n)$ be a corresponding iterated function system in $\mathbf{R}$; and let $A \subseteq \mathbf{R}$ be a nonempty measurable set. Suppose $\mathcal{L}(f_j[A] \cap f_k[A]) = 0$ for $j \neq k$, and*

$$A = \bigcup_{j=1}^{n} f_j[A].$$

If $0 < \mathcal{L}(A) < \infty$, then $s = 1$.

Number systems. Recall the situation from Section 1.6. Let b be a real number, and let D be a finite set of real numbers, including 0. We are interested in representing real numbers in the number system they define.

Write W for the set of "whole numbers"; that is, numbers of the form

$$\sum_{j=0}^{M} a_j b^j.$$

Write F for the set of "fractions"; that is numbers of the form

$$\sum_{j=-\infty}^{-1} a_j b^j.$$

We know (Proposition 3.1.10) that there is no number system that has a unique representation for every real number. So we will try to represent all real numbers, and arrange to have as few numbers as possible with more than one representation. One way to specify that the set with multiple representations is small is to require that it have Lebesgue measure 0.

If we analyze the size of the intersections

$$(F + w_1) \cap (F + w_2), \qquad w_1, w_2 \in W, \ w_1 \neq w_2,$$

then we will know about all numbers with multiple representations:

(5.1.19) EXERCISE. *The set of all numbers with multiple representations is a countable union of sets, each of which is similar to one of the sets* $(F + w_1) \cap (F + w_2)$, $w_1, w_2 \in W, w_1 \neq w_2$.

The set of all numbers that can be represented is $F + W$, a countable union of sets isometric to F. So if all real numbers can be represented, then $\mathcal{L}(F) > 0$. We know that F is a compact set, so also $\mathcal{L}(F) < \infty$. If the set of all real numbers with multiple representations has Lebesgue measure 0, then the sets $(F + w_1) \cap (F + w_2)$ have Lebesgue measure 0.

Suppose D has k elements. If F has positive Lebesgue measure, but the intersections $(F + w_1) \cap (F + w_2)$ have Lebesgue measure zero, then by Exercise 5.1.18, F has similarity dimension 1. But the similarity dimension is actually $\log k / \log |b|$. Therefore $|b| = k$.

5.2. METHOD I

We will need to discuss measures other than Lebesgue measure. The basics are contained in this section.

Measures and outer measures. A collection $\mathcal{F}$ of subsets of a set X is called a **σ-algebra** on X iff:

(1) $\varnothing, X \in \mathcal{F}$;
(2) if $A \in \mathcal{F}$, then $X \setminus A \in \mathcal{F}$;
(3) if $A_1, A_2, \cdots \in \mathcal{F}$, then $\bigcup_{i \in \mathbb{N}} A_i \in \mathcal{F}$.

Of course (by Theorem 5.1.14), the collection of all Lebesgue measurable subsets of $\mathbb{R}$ is a σ-algebra on $\mathbb{R}$. Combining the clauses of the definition will produce a few more rules: For example, if $A_1, A_2, \cdots \in \mathcal{F}$, then $\bigcap_{i \in \mathbb{N}} A_i \in \mathcal{F}$; if $A, B \in \mathcal{F}$ then $A \cap B, A \cup B, A \setminus B \in \mathcal{F}$.

(5.2.1) THEOREM. *Let X be a set, and let $\mathcal{D}$ be any set of subsets of X. Then there is a set $\mathcal{F}$ of subsets of X such that*

(1) $\mathcal{F}$ *is a σ-algebra on X;*
(2) $\mathcal{F} \supseteq \mathcal{D}$;
(3) *if $\mathcal{G}$ is any σ-algebra on X with $\mathcal{G} \supseteq \mathcal{D}$, then $\mathcal{G} \supseteq \mathcal{F}$.*

PROOF: First I claim that the intersection of any collection of σ-algebras on X is a σ-algebra. Let Γ be a collection of σ-algebras, and let $\mathcal{B} = \bigcap_{\mathcal{A} \in \Gamma} \mathcal{A}$ be the intersection. Then $\varnothing \in \mathcal{A}$ for all $\mathcal{A} \in \Gamma$, so $\varnothing \in \mathcal{B}$. Similarly $X \in \mathcal{B}$. If $A \in \mathcal{B}$, then $A \in \mathcal{A}$ for all $\mathcal{A} \in \Gamma$, so $X \setminus A \in \mathcal{A}$ for all $\mathcal{A} \in \Gamma$, and therefore $X \setminus A \in \mathcal{B}$. If $A_1, A_2, \cdots \in \mathcal{B}$, then each $A_n \in \mathcal{A}$ for all $\mathcal{A} \in \Gamma$, so $\bigcup_{n \in \mathbb{N}} A_n \in \mathcal{A}$ for all $\mathcal{A} \in \Gamma$ and therefore $\bigcup_{n \in \mathbb{N}} A_n \in \mathcal{B}$.

So suppose a set $\mathcal{D}$ of subsets of X is given. Let Γ be the collection of all σ-algebras $\mathcal{G}$ on X with $\mathcal{G} \supseteq \mathcal{D}$. (There is at least one such σ-algebra, namely the family of all subsets of X.) Then the intersection $\mathcal{F} = \bigcap_{\mathcal{G} \in \Gamma} \mathcal{G}$ is a σ-algebra on X. But clearly if $\mathcal{G}$ is any σ-algebra on X with $\mathcal{G} \supseteq \mathcal{D}$, then $\mathcal{G} \in \Gamma$, and therefore $\mathcal{G} \supseteq \mathcal{F}$. ☺

We say that $\mathcal{F}$ is the **least σ-algebra** containing $\mathcal{D}$, or the σ-algebra **generated** by $\mathcal{D}$. Let S be a metric space. A subset of S is called a **Borel set** iff it belongs to the σ-algebra on S generated by the open sets.

Let X be a set, and let $\mathcal{F}$ be a σ-algebra of subsets of X. A **measure** on $\mathcal{F}$ is a set function† $\mathcal{M}: \mathcal{F} \to [0, \infty]$ such that:

(1) $\mathcal{M}(\varnothing) = 0$;
(2) If $A_n \in \mathcal{F}$ is a disjoint sequence of sets, then

$$\mathcal{M}\left(\bigcup_{n \in \mathbb{N}} A_n\right) = \sum_{n=1}^{\infty} \mathcal{M}(A_n).$$

We call (2) **countable additivity**.

Let X be a set. An **outer measure** on X is a function $\overline{\mathcal{M}}$ defined on all subsets of X, with values in the nonnegative extended real numbers $[0, \infty]$, satisfying:

(1) $\overline{\mathcal{M}}(\varnothing) = 0$;
(2) if $A \subseteq B$, then $\overline{\mathcal{M}}(A) \leq \overline{\mathcal{M}}(B)$;
(3) $\overline{\mathcal{M}}\left(\bigcup_{n \in \mathbb{N}} A_n\right) \leq \sum_{n=1}^{\infty} \overline{\mathcal{M}}(A_n)$.

We call (3) **countable subadditivity**.

Generating outer measures. The Lebesgue outer measure was constructed in Section 5.1. The way in which the definition was formulated was not accidental. We will explore a general method for constructing outer measures, known as "method I". We begin with candidate values for the measures of some sets (like the lengths of the half-open intervals), and then attempt to produce an outer measure that is as large as possible, but no larger than the candidate values.

Let X be a set, and let $\mathcal{A}$ be a family of subsets of X that covers X. Let $C: \mathcal{A} \to [0, \infty]$ be any function. The theorem on construction of outer measures is as follows:

---

†A **set function** is a function whose domain is a family of sets.

(5.2.2) METHOD I THEOREM. *There is a unique outer measure $\overline{\mathcal{M}}$ on X such that*

(1) $\overline{\mathcal{M}}(A) \leq C(A)$ *for all $A \in \mathcal{A}$;*
(2) *if $\overline{\mathcal{N}}$ is any outer measure on X with $\overline{\mathcal{N}}(A) \leq C(A)$ for all $A \in \mathcal{A}$, then $\overline{\mathcal{N}}(B) \leq \overline{\mathcal{M}}(B)$ for all $B \subseteq X$.*

PROOF: The uniqueness is easy: if two outer measures satisfy (1) and (2), then each is $\leq$ the other, so they are equal.

For any subset B of X, define

$$(5.2.2a) \qquad \overline{\mathcal{M}}(B) = \inf \sum_{A \in \mathcal{D}} C(A),$$

where the infimum is over all countable covers $\mathcal{D}$ of B by sets of $\mathcal{A}$.

I claim first that $\overline{\mathcal{M}}$ is an outer measure. First, $\overline{\mathcal{M}}(\varnothing) = 0$, since the empty set is covered by the empty cover, and the empty sum has value 0. If $B \subseteq C$, then any cover of C is also a cover of B, so $\overline{\mathcal{M}}(B) \leq \overline{\mathcal{M}}(C)$. Let $B_1, B_2, \cdots$ be given. I must prove

$$\overline{\mathcal{M}}\left(\bigcup_{n \in \mathbf{N}} B_n\right) \leq \sum_{n=1}^{\infty} \overline{\mathcal{M}}(B_n).$$

If $\overline{\mathcal{M}}(B_n) = \infty$ for some n, then the inequality is clear. So suppose $\overline{\mathcal{M}}(B_n) < \infty$ for all n. Let $\varepsilon > 0$. For each n, choose a countable cover $\mathcal{D}_n$ of B_n by sets of $\mathcal{A}$ with

$$\sum_{A \in \mathcal{D}_n} C(A) \leq \overline{\mathcal{M}}(B_n) + 2^{-n}\varepsilon.$$

Now $\mathcal{D} = \bigcup_{n \in \mathbf{N}} \mathcal{D}_n$ is a countable cover of the union $\bigcup_{n \in \mathbf{N}} B_n$. Therefore

$$\overline{\mathcal{M}}\left(\bigcup_{n \in \mathbf{N}} B_n\right) \leq \sum_{A \in \mathcal{D}} C(A)$$

$$\leq \sum_{n=1}^{\infty} \sum_{A \in \mathcal{D}_n} C(A)$$

$$\leq \sum_{n=1}^{\infty} \overline{\mathcal{M}}(B_n) + \sum_{n=1}^{\infty} 2^{-n}\varepsilon$$

$$= \sum_{n=1}^{\infty} \overline{\mathcal{M}}(B_n) + \varepsilon.$$

Since ε was any positive number, we have

$$\overline{\mathcal{M}}\left(\bigcup_{n \in \mathbf{N}} B_n\right) \leq \sum_{n=1}^{\infty} \overline{\mathcal{M}}(B_n).$$

This completes the proof that $\overline{\mathcal{M}}$ is an outer measure.

Now we may check the two assertions of the theorem. For (1), note that for $A \in \mathcal{A}$, the singleton $\{A\}$ is a cover of A, so

$$\overline{\mathcal{M}}(A) \leq \sum_{B \in \{A\}} C(B) = C(A).$$

For (2), suppose that $\overline{\mathcal{N}}$ is any outer measure on X with $\overline{\mathcal{N}}(A) \leq C(A)$ for all $A \in \mathcal{A}$. Then for any countable cover $\mathcal{D}$ of a set B by elements of $\mathcal{A}$ we have

$$\sum_{A \in \mathcal{D}} C(A) \geq \sum_{A \in \mathcal{D}} \overline{\mathcal{N}}(A) \geq \overline{\mathcal{N}}\left(\bigcup_{A \in \mathcal{D}} A\right) \geq \overline{\mathcal{N}}(B).$$

Therefore $\overline{\mathcal{M}}(B) \geq \overline{\mathcal{N}}(B)$. ☺

When we say that an outer measure is to be constructed by **method I**, we are referring to this theorem. In practical terms, this means that the outer measure is defined by the formula (5.2.2a).

Measurable sets. Let $\overline{\mathcal{M}}$ be an outer measure on a set X. A set $A \subseteq X$ is $\overline{\mathcal{M}}$-measurable (in the sense of Carathéodory) iff $\overline{\mathcal{M}}(E) = \overline{\mathcal{M}}(E \cap A) + \overline{\mathcal{M}}(E \setminus A)$ for all sets $E \subseteq X$.

(5.2.3) THEOREM. *The collection $\mathcal{F}$ of $\overline{\mathcal{M}}$-measurable sets is a σ-algebra, and $\overline{\mathcal{M}}$ is countably additive on $\mathcal{F}$.*

PROOF: First, $\varnothing \in \mathcal{F}$ since for any E, we have $\overline{\mathcal{M}}(E \cap \varnothing) + \overline{\mathcal{M}}(E \setminus \varnothing) = \overline{\mathcal{M}}(\varnothing) + \overline{\mathcal{M}}(E) = \overline{\mathcal{M}}(E)$. It is also easy to see that a set A belongs to $\mathcal{F}$ if and only if its complement $X \setminus A$ does.

Suppose $A_j \in \mathcal{F}$ for $j = 1, 2, \cdots$. Let E be any test set. Then

$$\begin{aligned}
\overline{\mathcal{M}}(E) &= \overline{\mathcal{M}}(E \cap A_1) + \overline{\mathcal{M}}(E \setminus A_1) \\
&= \overline{\mathcal{M}}(E \cap A_1) + \overline{\mathcal{M}}((E \setminus A_1) \cap A_2) + \overline{\mathcal{M}}(E \setminus (A_1 \cup A_2)) \\
&= \cdots \\
&= \sum_{j=1}^{k} \overline{\mathcal{M}}\left(\left(E \setminus \bigcup_{i=1}^{j-1} A_i\right) \cap A_j\right) + \overline{\mathcal{M}}\left(E \setminus \bigcup_{j=1}^{k} A_j\right).
\end{aligned}$$

Hence

$$\overline{\mathcal{M}}(E) \geq \sum_{j=1}^{k} \overline{\mathcal{M}}\left(\left(E \setminus \bigcup_{i=1}^{j-1} A_i\right) \cap A_j\right) + \overline{\mathcal{M}}\left(E \setminus \bigcup_{j \in \mathbb{N}} A_j\right).$$

so (let $k \to \infty$)

$$\overline{\mathcal{M}}(E) \geq \sum_{j=1}^{\infty} \overline{\mathcal{M}}\left(\left(E \setminus \bigcup_{i=1}^{j-1} A_i\right) \cap A_j\right) + \overline{\mathcal{M}}\left(E \setminus \bigcup_{j \in \mathbb{N}} A_j\right).$$

But

$$E \cap \bigcup_{j \in \mathbf{N}} A_j = \bigcup_{j \in \mathbf{N}} \left(\left(E \setminus \bigcup_{i=1}^{j-1} A_i \right) \cap A_j \right),$$

so

$$\overline{\mathcal{M}}(E) \leq \overline{\mathcal{M}} \left(E \cap \bigcup_{j \in \mathbf{N}} A_j \right) + \overline{\mathcal{M}} \left(E \setminus \bigcup_{j \in \mathbf{N}} A_j \right)$$

$$\leq \sum_{j=1}^{\infty} \overline{\mathcal{M}} \left(\left(E \setminus \bigcup_{i=1}^{j-1} A_i \right) \cap A_j \right) + \overline{\mathcal{M}} \left(E \setminus \bigcup_{j \in \mathbf{N}} A_j \right)$$

$$\leq \overline{\mathcal{M}}(E).$$

Thus $\bigcup A_j \in \mathcal{F}$. This completes the proof that $\mathcal{F}$ is a σ-algebra.

Now if the sets $A_j \in \mathcal{F}$ are disjoint, we can let $E = \bigcup A_j$ in the previous computation, and we get

$$\overline{\mathcal{M}} \left(\bigcup_{j \in \mathbf{N}} A_j \right) = \sum_{j=1}^{\infty} \overline{\mathcal{M}}(A_j),$$

so $\overline{\mathcal{M}}$ is countably additive on $\mathcal{F}$. ☺

We will write simply $\mathcal{M}$ for the restriction of $\overline{\mathcal{M}}$ to the σ-algebra $\mathcal{F}$ of measurable sets. It is a measure on $\mathcal{F}$. Thus we see that we have a generalization of Lebesgue measure as constructed in Section 5.1.

(5.2.4) COROLLARY. *Every Borel set in $\mathbf{R}$ is Lebesgue measurable.*

PROOF: Open sets are measurable by Theorem 5.1.12. The collection of measurable sets is a σ-algebra by Theorem 5.1.14. ☺

Metric outer measure. Consider the following example of a method I outer measure on $\mathbf{R}$; the definition is very close to the definition used for Lebesgue measure. We begin with the collection $\mathcal{A} = \{ [a, b) : a < b \}$ of half-open intervals and the set function $C([a, b)) = \sqrt{b - a}$. Let $\overline{\mathcal{M}}$ be the corresponding method I outer measure. I claim that the interval $A = [0, 1]$ is not measurable.

Consider the measure of $[0, 1)$. Certainly the singleton $\{[0, 1)\}$ covers $[0, 1)$, so $\overline{\mathcal{M}}([0, 1)) \leq C([0, 1)) = 1$. If $[0, 1) \subseteq \bigcup_{i \in \mathbf{N}} [a_i, b_i)$, then by what we know about Lebesgue measure, we must have $\sum_{i=1}^{\infty}(b_i - a_i) \geq 1$. So we have also

$$\left(\sum_{i=1}^{\infty} \sqrt{b_i - a_i} \right)^2 = \sum_{i=1}^{\infty} \left(\sqrt{b_i - a_i} \right)^2 + 2 \sum_{i<j} \sqrt{b_i - a_i} \sqrt{b_j - a_j}$$

$$\geq \sum_{i=1}^{\infty}(b_i - a_i) \geq 1.$$

Therefore $\sum_{i=1}^{\infty} \sqrt{b_i - a_i} \geq 1$. This shows that $\overline{\mathcal{M}}([0,1)) \geq 1$. So we have $\overline{\mathcal{M}}([0,1)) = 1$.

Similarly $\overline{\mathcal{M}}([-1,0)) = 1$. The singleton $\{[-1,1)\}$ covers $[-1,1)$, so as before we have $\overline{\mathcal{M}}([-1,1)) \leq C([-1,1)) = \sqrt{2}$. So if $A = [0,1]$ and $E = [-1,1)$, we have

$$\overline{\mathcal{M}}(E \cap A) + \overline{\mathcal{M}}(E \setminus A) = 1 + 1 = 2 > \sqrt{2} \geq \overline{\mathcal{M}}(E).$$

This shows that $A = [0,1]$ is not measurable.

It is often desirable that the sets we work with be measurable sets. When we work with subsets of a metric space (as is common in this book), the sets are often open sets, closed sets, or sets constructed simply from open and closed sets. In particular, the sets are often Borel sets. There is a condition that will insure that all Borel sets are measurable sets.

Two sets A, B in a metric space have **positive separation** iff $\text{dist}(A, B) > 0$; that is, there is $r > 0$ with $\rho(x, y) \geq r$ for all $x \in A$ and $y \in B$. Let $\overline{\mathcal{M}}$ be an outer measure on a metric space S. We say that $\overline{\mathcal{M}}$ is a **metric outer measure** iff $\overline{\mathcal{M}}(A \cup B) = \overline{\mathcal{M}}(A) + \overline{\mathcal{M}}(B)$ for any pair A, B of sets with positive separation. (Theorem 5.1.6 shows that $\overline{\mathcal{L}}$ is a metric outer measure.) The measure $\mathcal{M}$ obtained by restricting a metric outer measure $\overline{\mathcal{M}}$ to its measurable sets will be called a **metric measure**.

The reason that metric outer measures are of interest is that open sets (and therefore all Borel sets) are measurable sets. Before I prove this, I formulate the lemma of Carathéodory.

(5.2.5) LEMMA. *Let $\overline{\mathcal{M}}$ be a metric outer measure on the metric space S. Let $A_1 \subseteq A_2 \subseteq \cdots$, and $A = \bigcup_{j \in \mathbb{N}} A_j$. Assume $\text{dist}(A_j, A \setminus A_{j+1}) > 0$ for all j. Then $\overline{\mathcal{M}}(A) = \lim_{j \to \infty} \overline{\mathcal{M}}(A_j)$.*

PROOF: For all j we have $\overline{\mathcal{M}}(A) \geq \overline{\mathcal{M}}(A_j)$, so $\overline{\mathcal{M}}(A) \geq \lim_{j \to \infty} \overline{\mathcal{M}}(A_j)$. (This inequality is true for any outer measure.) If $\lim_{j \to \infty} \overline{\mathcal{M}}(A_j) = \infty$, then the equation is true. So suppose $\lim_{j \to \infty} \overline{\mathcal{M}}(A_j) < \infty$.

Let $B_1 = A_1$ and $B_j = A_j \setminus A_{j-1}$ for $j \geq 2$. If $i \geq j+2$, then $B_j \subseteq A_j$ and $B_i \subseteq A \setminus A_{i-1} \subseteq A \setminus A_{j+1}$, so B_i and B_j have positive separation. So

$$\overline{\mathcal{M}} \left(\bigcup_{k=1}^{m} B_{2k-1} \right) = \sum_{k=1}^{m} \overline{\mathcal{M}}(B_{2k-1})$$

$$\overline{\mathcal{M}} \left(\bigcup_{k=1}^{m} B_{2k} \right) = \sum_{k=1}^{m} \overline{\mathcal{M}}(B_{2k}).$$

Since $\lim_{j \to \infty} \overline{\mathcal{M}}(A_j) < \infty$, both of these converge (as $m \to \infty$). So

$$\overline{\mathcal{M}}(A) = \overline{\mathcal{M}}\left(\bigcup_{j \in \mathbf{N}} A_j\right) = \overline{\mathcal{M}}\left(A_j \cup \bigcup_{k \geq j+1} B_k\right)$$

$$\leq \overline{\mathcal{M}}(A_j) + \sum_{k=j+1}^{\infty} \overline{\mathcal{M}}(B_k)$$

$$\leq \lim_{i \to \infty} \overline{\mathcal{M}}(A_i) + \sum_{k=j+1}^{\infty} \overline{\mathcal{M}}(B_k).$$

Now as $j \to \infty$, the tail of a convergent series goes to 0, so we get

$$\overline{\mathcal{M}}(A) \leq \lim_{i \to \infty} \overline{\mathcal{M}}(A_i). \ \copyright$$

(5.2.6) THEOREM. *Let $\overline{\mathcal{M}}$ be a metric outer measure on a metric space S. Then every Borel subset of S is $\overline{\mathcal{M}}$-measurable.*

PROOF: Since the σ-algebra of Borel sets is the σ-algebra generated by the closed sets, and since the collection $\mathcal{F}$ of measurable sets is a σ-algebra, it is enough to show that every closed set F is measurable. Let A be any test set. I must show that $\overline{\mathcal{M}}(A) \geq \overline{\mathcal{M}}(A \cap F) + \overline{\mathcal{M}}(A \setminus F)$, since the opposite inequality is true for any outer measure.

Let $A_j = \{\, x \in A : \operatorname{dist}(x, F) \geq 1/j \,\}$. Then $\operatorname{dist}(A_j, F \cap A) \geq 1/j$, so

(5.2.6a) $$\overline{\mathcal{M}}(A \cap F) + \overline{\mathcal{M}}(A_j) = \overline{\mathcal{M}}\big((A \cap F) \cup A_j\big) \leq \overline{\mathcal{M}}(A).$$

Now since F is closed, F contains all points of distance 0 from F, so $A \setminus F = \bigcup_{j \in \mathbf{N}} A_j$. We check the condition of the lemma: If $x \in \big(A \setminus (F \cup A_{j+1})\big)$, then there exists $z \in F$ with $\rho(x, z) < 1/(j+1)$. If $y \in A_j$, then

$$\rho(x, y) \geq \rho(y, z) - \rho(x, z) > \frac{1}{j} - \frac{1}{j+1}.$$

Thus

$$\operatorname{dist}\big(A \setminus (F \cup A_{j+1}), A_j\big) \geq \frac{1}{j} - \frac{1}{j+1} > 0.$$

Therefore, applying the lemma, we get $\overline{\mathcal{M}}(A \setminus F) \leq \lim_{j \to \infty} \overline{\mathcal{M}}(A_j)$. Taking the limit in (5.2.6a), we get $\overline{\mathcal{M}}(A \cap F) + \overline{\mathcal{M}}(A \setminus F) \leq \overline{\mathcal{M}}(A)$, which completes the proof. $\copyright$

(5.2.7) EXERCISE. *Let $\mathcal{M}$ be a finite metric measure on a compact metric space S. Let $E \subseteq S$ be a measurable set. For any $\varepsilon > 0$, there exist a compact set K and an open set U with $K \subseteq E \subseteq U$ and $\mathcal{M}(U \setminus K) < \varepsilon$.*

5.3. Two-Dimensional Lebesgue Measure

We next define two-dimensional Lebesgue measure. This is a measure defined for subsets of the plane $\mathbf{R}^2$.

A **rectangle** in $\mathbf{R}^2$ is a set R of the form

$$R = [a, b) \times [c, d)$$
$$= \{ (x, y) \in \mathbf{R}^2 : a \leq x < b, c \leq y < d \}$$

for some $a \leq b$ and $c \leq d$. The **area** of this rectangle is $A(R) = (b-a)(d-c)$, as usual. In particular, if $a = b$ or $c = d$, we see that $A(\varnothing) = 0$. **Two-dimensional Lebesgue outer measure** is the outer measure $\overline{\mathcal{L}}^2$ on $\mathbf{R}^2$ defined by method I from this function A.

(5.3.1) Theorem. *Two-dimensional Lebesgue outer measure $\overline{\mathcal{L}}^2$ is a metric outer measure.*

PROOF: Suppose A and B are sets with positive separation. Since $\overline{\mathcal{L}}^2$ is an outer measure, we have $\overline{\mathcal{L}}^2(A \cup B) \leq \overline{\mathcal{L}}^2(A) + \overline{\mathcal{L}}^2(B)$. So I must prove the opposite inequality.

Let $\mathcal{D}$ be a cover of $A \cup B$ by rectangles. Now a rectangle $R = [a, b) \times [c, d)$ can be written as a union of the four rectangles

$$[a, (a+b)/2) \times [c, (c+d)/2)$$
$$[a, (a+b)/2) \times [(c+d)/2, d)$$
$$[(a+b)/2, b) \times [c, (c+d)/2)$$
$$[(a+b)/2, b) \times [(c+d)/2, d),$$

and the area of the large rectangle is the sum of the areas of the four small rectangles. So the sum

$$\sum_{R \in \mathcal{D}} A(R)$$

is unchanged when we replace one of the rectangles by its four parts. Applying this repeatedly, we may assume that the diameters of the rectangles in $\mathcal{D}$ are all smaller than $\mathrm{dist}(A, B)$. Then no rectangle of $\mathcal{D}$ intersects both A and B. So $\mathcal{D}$ is a disjoint union of two families, $\mathcal{A}$ and $\mathcal{B}$, where $\mathcal{A}$ covers A and $\mathcal{B}$ covers B. But then

$$\sum_{R \in \mathcal{D}} A(R) = \sum_{R \in \mathcal{A}} A(R) + \sum_{R \in \mathcal{B}} A(R) \geq \overline{\mathcal{L}}^2(A) + \overline{\mathcal{L}}^2(B).$$

So we conclude that $\overline{\mathcal{L}}^2(A \cup B) \geq \overline{\mathcal{L}}^2(A) + \overline{\mathcal{L}}^2(B)$. ☺

The sets that are measurable in the sense of Carathéodory for $\overline{\mathcal{L}}^2$ are again called the **Lebesgue measurable sets**; the restriction of $\overline{\mathcal{L}}^2$ to this σ-algebra is called **two-dimensional Lebesgue measure**. We will write $\mathcal{L}^2$ for two-dimensional Lebesgue measure.

The fact that two-dimensional Lebesgue measure is not identically zero is left to you:

(5.3.2) EXERCISE. *If $a < b$ and $c < d$, then a rectangle of the form $R = [a, b) \times [c, d)$ is Lebesgue measurable and $\mathcal{L}^2(R) = (b - a)(d - c)$.*

Now that we know that the Lebesgue measure of a square is what it should be, the usual scheme of approximating an area with a lot of little squares will show that the usual sets of Euclidean plane geometry have two-dimensional Lebesgue measure equal to their areas. In particular, a rectangle with sides not parallel to the coordinate axes has the right area. This should be enough to prove:

(5.3.3) EXERCISE. *Let $f: \mathbf{R}^2 \to \mathbf{R}^2$ be a similarity with ratio r. If $A \subseteq \mathbf{R}^2$ is Lebesgue measurable, then so is $f[A]$, and $\mathcal{L}^2(f[A]) = r^2 \mathcal{L}^2(A)$.*

This will tell us something about the similarity dimension of a set $A \subseteq \mathbf{R}^2$, as in the one-dimensional case (Exercise 5.1.18).

(5.3.4) EXERCISE. *Let $(r_1, r_2, \cdots, r_n)$ be a contracting ratio list; let s be its dimension; let $(f_1, f_2, \cdots f_n)$ is a corresponding iterated function system in $\mathbf{R}^2$; and let $A \subseteq \mathbf{R}^2$ be a nonempty Borel set. Suppose $\mathcal{L}^2(f_j[A] \cap f_k[A]) = 0$ for $j \neq k$, and $A = \bigcup_{j=1}^n f_j[A]$. If $0 < \mathcal{L}^2(A) < \infty$, then $s = 2$.*

This result can be used for complex number systems in the same way as the corresponding result was used for real number systems in (5.1.18).

(5.3.5) EXERCISE. *Let b be a complex number, and let D be a finite set of complex numbers, including 0. Suppose D has k elements. Suppose every complex number can be represented in the number system defined by base b and digit set D, and the set of complex numbers with multiple representations has two-dimensional Lebesgue measure 0. What does this mean about the relationship between b and k?*

Higher dimensions. Let d be a positive integer. In d-dimensional Euclidean space $\mathbf{R}^d$, we will consider **hyper-rectangles** of the form

$$R = [a_1, b_1) \times [a_2, b_2) \times \cdots \times [a_d, b_d),$$

where $a_j < b_j$ for all j. The "hyper-volume" of this hyper-rectangle R is

$$A(R) = \prod_{j=1}^{d} (b_j - a_j).$$

We define d-**dimensional Lebesgue outer measure** to be the method I outer measure defined from this set function A. We define d-**dimensional Lebesgue measure** to be the restriction to the measurable subsets. As before, we use the notation $\overline{\mathcal{L}}^d$ and $\mathcal{L}^d$.

(5.3.6) EXERCISE. *The outer measure $\overline{\mathcal{L}}^d$ is a metric outer measure. If*

$$R = [a_1, b_1) \times [a_2, b_2) \times \cdots \times [a_d, b_d),$$

where $a_j \leq b_j$ for all j, then

$$\mathcal{L}^d(R) = \prod_{j=1}^{d} (b_j - a_j).$$

5.4. METHOD II

We have seen that method I may fail to yield a measure where open sets are measurable. There is a related construction, called "method II" that will overcome this difficulty.

(5.4.1) EXERCISE. *Let $\mathcal{A} \subseteq \mathcal{B}$ be two covers of X, and let $C: \mathcal{B} \to [0, \infty]$ be a set function. If $\overline{\mathcal{M}}$ is the method I outer measure defined by C and $\mathcal{A}$, and if $\overline{\mathcal{N}}$ is the method I outer measure defined by C and $\mathcal{B}$, then $\overline{\mathcal{M}}(A) \geq \overline{\mathcal{N}}(A)$ for all $A \subseteq X$.*

Let $\mathcal{A}$ be a family of subsets of a metric space S, and suppose, for every $x \in S$ and $\varepsilon > 0$, there exists $A \in \mathcal{A}$ with $x \in A$ and diam $A \leq \varepsilon$. Suppose $C: \mathcal{A} \to [0, \infty]$ is a given function. An outer measure will be constructed based on this data. For each $\varepsilon > 0$, let

$$\mathcal{A}_\varepsilon = \{ A \in \mathcal{A} : \text{diam } A \leq \varepsilon \}.$$

Let $\overline{\mathcal{M}}_\varepsilon$ be the method I outer measure determined by C using the family $\mathcal{A}_\varepsilon$. Then by Exercise 5.4.1, for a given set E, when ε decreases, $\overline{\mathcal{M}}_\varepsilon(E)$ increases. Define

$$\overline{\mathcal{M}}(E) = \lim_{\varepsilon \to 0} \overline{\mathcal{M}}_\varepsilon(E) = \sup_{\varepsilon > 0} \overline{\mathcal{M}}_\varepsilon(E).$$

It is easily verified that $\overline{\mathcal{M}}$ is an outer measure. As usual, we will write $\mathcal{M}$ for the restriction to the measurable sets. This construction of an outer measure $\overline{\mathcal{M}}$ from a set function C (and a measure $\mathcal{M}$ from $\overline{\mathcal{M}}$) is called **method II**. It is more complicated than method I, but (unlike method I) it insures that Borel sets are measurable:

(5.4.2) THEOREM. *The set function $\overline{\mathcal{M}}$ defined by method II is a metric outer measure.*

PROOF: Let $A, B \subseteq S$ with dist$(A, B) > 0$. Since $\overline{\mathcal{M}}$ is an outer measure, we have $\overline{\mathcal{M}}(A \cup B) \leq \overline{\mathcal{M}}(A) + \overline{\mathcal{M}}(B)$. So I must prove the opposite inequality.

Let $\varepsilon > 0$ so small that $\varepsilon < $ dist(A, B). Let $\mathcal{D}$ be any countable cover of $A \cup B$ by sets of $\mathcal{A}_\varepsilon$. The sets $D \in \mathcal{D}$ have diameter less than dist(A, B), so such a set D intersects at most one of the sets A, B. Therefore, $\mathcal{D}$ may be divided into two disjoint collections, $\mathcal{D}_1$ and $\mathcal{D}_2$, where $\mathcal{D}_1$ covers A and $\mathcal{D}_2$ covers B. Then

$$\sum_{D \in \mathcal{D}} C(D) = \sum_{D \in \mathcal{D}_1} C(D) + \sum_{D \in \mathcal{D}_2} C(D) \geq \overline{\mathcal{M}}_\varepsilon(A) + \overline{\mathcal{M}}_\varepsilon(B).$$

Now we may take the infimum over all covers, and conclude $\overline{\mathcal{M}}_\varepsilon(A \cup B) \geq \overline{\mathcal{M}}_\varepsilon(A) + \overline{\mathcal{M}}_\varepsilon(B)$. Then we may take the limit as $\varepsilon \to 0$ to conclude $\overline{\mathcal{M}}(A \cup B) \geq \overline{\mathcal{M}}(A) + \overline{\mathcal{M}}(B)$. ☺

5.5. MEASURES FOR STRINGS

One of the useful ways we will employ the material of this chapter is by defining measures. Sometimes (for example Lebesgue measure or Hausdorff measure) we will define a measure on subsets of Euclidean space $\mathbf{R}^d$. But also measures will be defined on our string models and path models.

An example. We will consider an easy example before we attack the more general case.

Begin with the two-letter alphabet $E = \{\mathbf{0}, \mathbf{1}\}$. Consider, as usual, the metric space $E^{(\omega)}$ of infinite strings with metric $\rho_{1/2}$. We will construct a measure on $E^{(\omega)}$. We begin with the family of "basic open sets":

$$\mathcal{A} = \left\{ [\alpha] : \alpha \in E^{(*)} \right\}$$

together with the set function $C \colon \mathcal{A} \to [0, \infty)$ defined by

$$C([\alpha]) = \left(\frac{1}{2} \right)^{|\alpha|} .$$

(Recall the notation $|\alpha|$ for the length of the string α.)

(5.5.1) PROPOSITION. *The method I outer measure* $\overline{\mathcal{M}}_{1/2}$ *constructed using this function C is a metric outer measure and satisfies* $\overline{\mathcal{M}}_{1/2}([\alpha]) = C([\alpha])$ *for all* $\alpha \in E^{(*)}$.

PROOF: Write $\mathcal{A} = \left\{ [\alpha] : \alpha \in E^{(*)} \right\}$, $\mathcal{A}_\varepsilon = \{ D \in \mathcal{A} : \operatorname{diam} D \le \varepsilon \}$. Let $\overline{\mathcal{N}}_\varepsilon$ be the method I measure defined by the set function C restricted to $\mathcal{A}_\varepsilon$. If $D \in \mathcal{A}_\varepsilon$, then of course $C(D) \ge \overline{\mathcal{M}}_{1/2}(D)$, so by the Method I theorem,

$$\overline{\mathcal{N}}_\varepsilon(A) \ge \overline{\mathcal{M}}_{1/2}(A)$$

for all A. Therefore the method II measure $\overline{\mathcal{N}}$ defined by

$$\overline{\mathcal{N}}(A) = \lim_{\varepsilon \to 0} \overline{\mathcal{N}}_\varepsilon(A)$$

satisfies $\overline{\mathcal{N}}(A) \ge \overline{\mathcal{M}}_{1/2}(A)$.

Note that, for any $\alpha \in E^{(*)}$, if k is the length $|\alpha|$, then we have

$$C([\alpha]) = \left(\frac{1}{2} \right)^k$$
$$= \left(\frac{1}{2} \right)^{k+1} + \left(\frac{1}{2} \right)^{k+1}$$
$$= C([\alpha\mathbf{0}]) + C([\alpha\mathbf{1}]).$$

Applying this repeatedly, we see that, for any $\varepsilon > 0$, any set $D \in \mathcal{A}$ is a finite disjoint union $D_1 \cup D_2 \cup \cdots \cup D_n$ of sets in $\mathcal{A}_\varepsilon$ with $C(D) = \sum C(D_i)$. This means that $\overline{\mathcal{N}}_\varepsilon(D) \le C(D)$, so by the Method I theorem, $\overline{\mathcal{N}}_\varepsilon(A) \le \overline{\mathcal{M}}_{1/2}(A)$ for all sets A. So $\overline{\mathcal{N}}(A) \le \overline{\mathcal{M}}_{1/2}(A)$

Therefore $\overline{\mathcal{M}}_{1/2} = \overline{\mathcal{N}}$ is a method II outer measure, so it is a metric outer measure. ☺

(5.5.2) EXERCISE. *Let* $h\colon E^{(\omega)} \to \mathbf{R}$ *be the "base 2" model map defined on page 13. If* $A \subseteq E^{(\omega)}$ *is a Borel set, then* $\mathcal{M}_{1/2}(A) = \mathcal{L}(h[A])$.

Measures on string spaces. Let E be a finite alphabet with at least two letters. Consider the space $E^{(\omega)}$ of infinite strings. This space is a metric space for many different metrics ρ. But all of the metrics constructed according to the scheme in Proposition 2.5.5 produce the same open sets. A countable base for the open sets is $\{\,[\alpha] : \alpha \in E^{(*)}\,\}$. Also $E^{(\omega)}$ is a compact ultrametric space. An important feature of all these metrics is $\lim_{k\to\infty} \operatorname{diam}[\sigma{\restriction}k] = 0$ for each $\sigma \in E^{(\omega)}$.

(5.5.3) EXERCISE. *It follows that* $\lim_{k\to\infty} \left(\sup\{\,\operatorname{diam}[\alpha] : \alpha \in E^{(k)}\,\}\right) = 0$.

Suppose a non-negative number w_α is given for each finite string α. Under what conditions is there a metric outer measure $\overline{\mathcal{M}}$ on $E^{(\omega)}$ with $\overline{\mathcal{M}}([\alpha]) = w_\alpha$ for all α? Since it is a metric outer measure, the open sets $[\alpha]$ are measurable, so $\overline{\mathcal{M}}$ is additive on them. Now the set $[\alpha]$ is the disjoint union of the sets $[\beta]$ as β ranges over the children of α (that is, $\beta = \alpha e$ for $e \in E$).

(5.5.4) THEOREM. *Suppose the non-negative numbers* w_α *satisfy*

$$w_\alpha = \sum_{e \in E} w_{\alpha e}$$

for $\alpha \in E^{(*)}$. *Then the method I outer measure defined by the set function* $C([\alpha]) = w_\alpha$ *is a metric outer measure* $\overline{\mathcal{M}}$ *on* $E^{(\omega)}$ *with* $\overline{\mathcal{M}}([\alpha]) = w_\alpha$.

PROOF: Write $\mathcal{A} = \{\,[\alpha] : \alpha \in E^{(*)}\,\}$, $\mathcal{A}_\varepsilon = \{\,D \in \mathcal{A} : \operatorname{diam} D \le \varepsilon\,\}$. Let $\overline{\mathcal{N}}_\varepsilon$ be the method I measure defined by the set function C restricted to $\mathcal{A}_\varepsilon$. If $D \in \mathcal{A}_\varepsilon$, then of course $C(D) \ge \overline{\mathcal{M}}(D)$, so by the Method I theorem,

$$\overline{\mathcal{N}}_\varepsilon(A) \ge \overline{\mathcal{M}}(A)$$

for all A. Therefore the method II measure $\overline{\mathcal{N}}$ defined by

$$\overline{\mathcal{N}}(A) = \lim_{\varepsilon \to 0} \overline{\mathcal{N}}_\varepsilon(A)$$

satisfies $\overline{\mathcal{N}}(A) \ge \overline{\mathcal{M}}(A)$.

Note that, for any $\alpha \in E^{(*)}$, we have

$$C([\alpha]) = w_\alpha = \sum_{e \in E} w_{\alpha e} = \sum_{e \in E} C([\alpha e]).$$

Applying this repeatedly, together with Exercise 5.5.3, we see that, for any $\varepsilon > 0$, any set $D \in \mathcal{A}$ is a finite disjoint union $D_1 \cup D_2 \cup \cdots \cup D_n$ of sets in $\mathcal{A}_\varepsilon$ with $C(D) = \sum C(D_i)$. This means that $\overline{\mathcal{N}}_\varepsilon(D) \le C(D)$, so by the Method I theorem, $\overline{\mathcal{N}}_\varepsilon(A) \le \overline{\mathcal{M}}(A)$ for all sets A. So $\overline{\mathcal{N}}(A) \le \overline{\mathcal{M}}(A)$.

Therefore $\overline{\mathcal{M}} = \overline{\mathcal{N}}$ is a method II outer measure. So we may conclude that it is a metric outer measure. ☺

How should we formulate the corresponding theorem for the path spaces $E_v^{(\omega)}$ defined by a directed multigraph (V, E, i, t)? We will define measures for each of the spaces $E_v^{(\omega)}$. We only need to define one of them at a time.

Fix a vertex v. Suppose nonnegative numbers w_α are given, one for each $\alpha \in E_v^{(*)}$. They should (of course) satisfy

$$w_\alpha = \sum_{i(e)=t(\alpha)} w_{\alpha e}$$

for $\alpha \in E_v^{(*)}$. Note that this has consequences for the troublesome exceptional cases that came up when we were defining the metric. If α has no children, then (interpreting an empty sum as 0), we see that $w_\alpha = 0$. Similarly, if α has only one child β, then $w_\alpha = w_\beta$.

(5.5.5) EXERCISE. *Suppose the non-negative numbers w_α satisfy*

$$w_\alpha = \sum_{i(e)=t(\alpha)} w_{\alpha e}$$

for $\alpha \in E_v^{(*)}$. Then the method I outer measure defined by the set function $C([\alpha]) = w_\alpha$ is a metric outer measure $\overline{\mathcal{M}}$ on $E_v^{(\omega)}$ with $\overline{\mathcal{M}}([\alpha]) = w_\alpha$.

*5.6. REMARKS

Henri Lebesgue's measure and integration theory dates from about 1900. It is one of the cornerstones of twentieth century mathematics. I have not discussed integration at all, in order to reduce the amount of material to its bare minimum. The more abstract measure theory was developed by many others, such as Constantin Carathéodory, during the early 1900's. Theorem 5.5.4 on the existence of measures on the string spaces is due essentially to A. N. Kolmogorov.

The σ-algebra $\mathcal{F}$ generated by a family $\mathcal{D}$ of sets (as in Theorem 5.2.1) is a complicated object to describe constructively. The proof given for 5.2.1 has the advantage of not requiring such a constructive description. Certainly $\mathcal{F}$ contains all countable unions

$$\bigcup_{i \in \mathbf{N}} D_i$$

with $D_i \in \mathcal{D}$; it contains complements of those unions; it contains countable intersections

$$\bigcap_{i \in \mathbf{N}} E_i$$

---

*An optional section.

where each E_i is either a countable union or a complement of a countable union. But that may not be everything in $\mathcal{F}$. (See, for example, the proof of Theorem (10.23) in [29].)

Exercise 5.3.5: $k = |b|^2$.

Exercise 5.5.2. Both measures are method I measures; use the Method I theorem twice, once to prove an inequality in each direction.

O, wiste a man how manye maladyes
Folwen of excesse and of glotonyes
He wolde been the moore mesurable
—G. Chaucer, *The Pardoner's Tale*

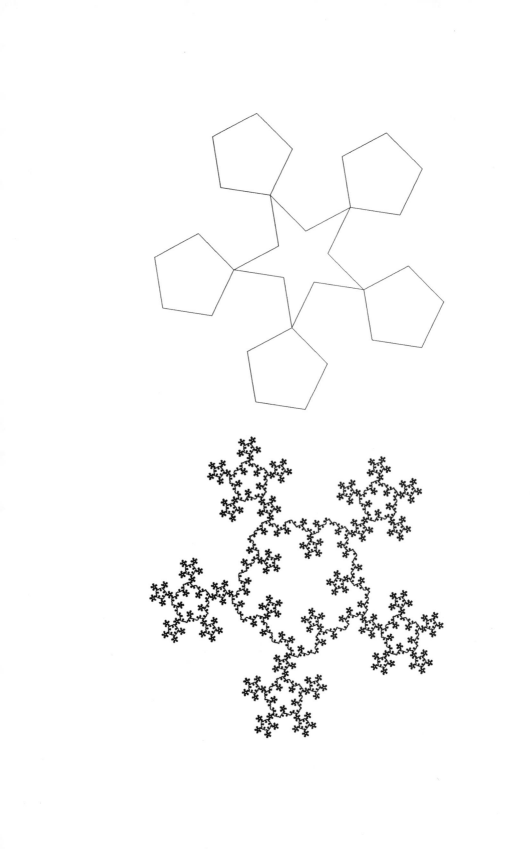

6

Hausdorff Dimension

Next we come to the "Hausdorff dimension". This is the dimension singled out by Mandelbrot when he defined "fractal". It is perhaps a bit more difficult to define than some of the other kinds of dimension that have been (and will be) considered. But it is also the most useful of the fractal dimensions.

6.1. Hausdorff Measure

The definition. Let S be a metric space. Consider a positive real number s, the candidate for the dimension. The **s-dimensional Hausdorff outer measure** is the method II outer measure defined from the set function $C(A) = (\text{diam } A)^s$. It is written $\overline{\mathcal{H}}^s$. The restriction to the measurable sets is called **s-dimensional Hausdorff measure**, and written $\mathcal{H}^s$. Since $\overline{\mathcal{H}}^s$ is constructed by method II, it is a metric outer measure. So, in particular, all Borel sets are measurable (in particular, all open sets, closed sets, compact sets).

The Method I theorem gives a more explicit formulation of the definition. A family $\mathcal{A}$ of subsets of S is called a **countable cover** of a set F iff

$$F \subseteq \bigcup_{A \in \mathcal{A}} A,$$

and $\mathcal{A}$ is a countable (possibly even finite) family of sets. Let ε be a positive number (presumably very small). The cover $\mathcal{A}$ is an **ε-cover** iff diam $A \leq \varepsilon$ for all $A \in \mathcal{A}$. Define

$$\overline{\mathcal{H}}^s_\varepsilon(F) = \inf \sum_{A \in \mathcal{A}} (\text{diam } A)^s,$$

where the infimum is over all countable ε-covers $\mathcal{A}$ of the set F. (By convention, $\inf \varnothing = \infty$.)

A computation shows that when ε gets smaller, $\overline{\mathcal{H}}^s_\varepsilon(F)$ gets larger. Finally:

$$\overline{\mathcal{H}}^s(F) = \lim_{\varepsilon \to 0} \overline{\mathcal{H}}^s_\varepsilon(F) = \sup_{\varepsilon > 0} \overline{\mathcal{H}}^s_\varepsilon(F)$$

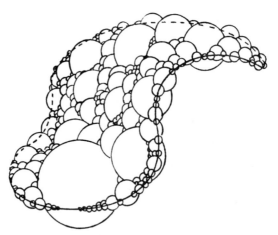

Figure 6.1.1. The Hausdorff measure (area) of a piece of surface A is approximated by the cross-sections of little balls which cover it. (From [43])

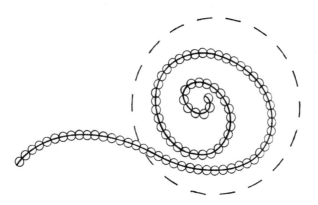

Figure 6.1.2. One must cover by *small* sets to compute length accurately. Here the length of the spiral is well-estimated by the sum of the diameters of the tiny balls, but grossly under-estimated by the diameter of the huge ball. (From [43])

is the **s-dimensional Hausdorff outer measure** of the set F. Figures 6.1.1 and 6.1.2 illustrate some of the ideas behind the definition.

(6.1.3) EXERCISE. *If F is a finite set, then $\mathcal{H}^s(F) = 0$ for all $s > 0$.*

(6.1.4) THEOREM. *In the metric space $\mathbb{R}$, the one-dimensional Hausdorff measure $\mathcal{H}^1$ coincides with the Lebesgue measure $\mathcal{L}$.*

PROOF: If $A \subseteq \mathbf{R}$ has finite diameter r, then $\sup A - \inf A = r$, so A is contained in a closed interval I with length r, and $\overline{\mathcal{L}}(A) \leq \overline{\mathcal{L}}(I) = r$. But by the Method I theorem (5.2.2), $\overline{\mathcal{H}}_\varepsilon^1$ is the largest outer measure $\overline{\mathcal{M}}$ satisfying $\overline{\mathcal{M}}(A) \leq \operatorname{diam} A$ for all sets A with diameter less than ε. So $\overline{\mathcal{H}}_\varepsilon^1(F) \geq \overline{\mathcal{L}}(F)$ for all F. Therefore $\overline{\mathcal{H}}^1(F) \geq \overline{\mathcal{L}}(F)$.

Now if $[a, b)$ is a half-open interval and $\varepsilon > 0$, we may find points $a = x_0 < x_1 < \cdots < x_n = b$ with $x_j - x_{j-1} < \varepsilon$ for all j. Then $[a, b)$ is covered by the countable collection $\{ [x_{j-1}, x_j] : 1 \leq j \leq n \}$, and

$$\sum_{j=1}^{n} \operatorname{diam}[x_{j-1}, x_j] = \sum_{j=1}^{n} (x_j - x_{j-1}) = b - a.$$

Therefore $\overline{\mathcal{H}}_\varepsilon^1\big([a, b)\big) \leq b - a$. But by the Method I theorem, $\overline{\mathcal{L}}$ is the largest outer measure satisfying $\overline{\mathcal{L}}\big([a, b)\big) \leq b - a$ for all half-open intervals $[a, b)$. Therefore $\overline{\mathcal{L}}(F) \geq \overline{\mathcal{H}}^1(F)$. for all F.

The two outer measures $\overline{\mathcal{L}}$ and $\overline{\mathcal{H}}^1$ coincide. The measurable sets in each case are given by the criterion of Carathéodory, so the measures $\mathcal{L}$ and $\mathcal{H}^1$ also coincide. ☺

Hausdorff dimension. How does the Hausdorff measure $\mathcal{H}^s(F)$ behave as a function of s for a given set F? An easy calculation shows that when s increases, $\mathcal{H}^s(F)$ decreases. But much more is true.

(6.1.5) THEOREM. *Let F be a Borel set. Let $0 < s < t$. If $\mathcal{H}^s(F) < \infty$, then $\mathcal{H}^t(F) = 0$. If $\mathcal{H}^t(F) > 0$, then $\mathcal{H}^s(F) = \infty$.*

PROOF: If $\operatorname{diam} A \leq \varepsilon$, then

$$\overline{\mathcal{H}}_\varepsilon^t(A) \leq (\operatorname{diam} A)^t \leq \varepsilon^{t-s}(\operatorname{diam} A)^s.$$

Therefore by the Method I theorem, $\overline{\mathcal{H}}_\varepsilon^t(F) \leq \varepsilon^{t-s}\overline{\mathcal{H}}_\varepsilon^s(F)$ for all F. Now if $\mathcal{H}^s(F) < \infty$, then $\mathcal{H}^t(F) \leq \lim_{\varepsilon \to 0} \varepsilon^{t-s}\,\overline{\mathcal{H}}_\varepsilon^s(F) = 0 \cdot \mathcal{H}^s(F) = 0$. The second assertion is the contrapositive. ☺

This means that, for a given set F, there is a unique "critical value" $s_0 \in [0, \infty]$ such that:

$$\mathcal{H}^s(F) = \infty \qquad \text{for all } s < s_0;$$
$$\mathcal{H}^s(F) = 0 \qquad \text{for all } s > s_0.$$

This value s_0 is called the **Hausdorff dimension** of the set F. We will write $s_0 = \dim F$. Of course, it is possible that $\mathcal{H}^s(F) = 0$ for all $s > 0$; in that case $\dim F = 0$. In the same way, it is possible that $\mathcal{H}^s(F) = \infty$ for all s; in that case $\dim F = \infty$.

There are variants in the definition of the Hausdorff dimension that are sometimes useful. Since the closure of a set has the same diameter as the set itself, we may use only closed sets in the covers $\mathcal{A}$. If A is any set, it is contained in an open set with diameter as close as I like to the diameter of A. So using only open sets in the covers $\mathcal{A}$ does not change the Hausdorff dimension. Any set of diameter r is contained in a closed ball of radius r (and diameter $\leq 2r$). So the s-dimensional Hausdorff measure is altered by at most a factor of 2^s if covers $\mathcal{A}$ by open balls are used; the dimension is not altered at all. If a set is compact, then every open cover has a finite subcover, so to compute the Hausdorff dimension of a compact set, we may use finite covers $\mathcal{A}$. If we replace a set in a cover $\mathcal{A}$ of F by a subset of itself, so that the result is still a cover of F, the sum

$$\sum_{A \in \mathcal{A}} (\operatorname{diam} A)^s$$

only becomes smaller. So if $F \subseteq T \subseteq S$, the value of $\overline{\mathcal{H}}^s_\varepsilon(F)$ when F is considered to be a subset of T is the same as when F is considered to be a subset of S. In particular, we may assume the sets used in the covers $\mathcal{A}$ of the set F are subsets of F.

(6.1.6) THEOREM. *Let A, B be Borel sets.*
- (1) *If $A \subseteq B$, then $\dim A \leq \dim B$.*
- (2) *$\dim(A \cup B) = \max\{\dim A, \dim B\}$.*

PROOF: (1) Suppose $A \subseteq B$. If $s > \dim B$, then $\mathcal{H}^s(A) \leq \mathcal{H}^s(B) = 0$. Therefore $\dim A \leq s$. This is true for all $s > \dim B$, so $\dim A \leq \dim B$.

(2) Let $s > \max\{\dim A, \dim B\}$. Then $s > \dim A$, so $\mathcal{H}^s(A) = 0$. Similarly, $\mathcal{H}^s(B) = 0$. Then $\mathcal{H}^s(A \cup B) \leq \mathcal{H}^s(A) + \mathcal{H}^s(B) = 0$. Therefore $\dim(A \cup B) \leq s$. This is true for all $s > \max\{\dim A, \dim B\}$, so we have $\dim(A \cup B) \leq \max\{\dim A, \dim B\}$. By (1), $\dim(A \cup B) \geq \max\{\dim A, \dim B\}$. ☺

(6.1.7) EXERCISE. *Suppose $A_1, A_2, \cdots$ are Borel sets. Is it true that*

$$\dim \bigcup_{k \in \mathbb{N}} A_k = \sup_k \dim A_k?$$

(6.1.8) THEOREM. *Let $f \colon S \to T$ be a similarity with ratio $r > 0$, let s be a positive real number, and let $F \subseteq S$ be a Borel set. Then $\mathcal{H}^s(f[F]) = r^s \mathcal{H}^s(F)$. So $\dim f[F] = \dim F$.*

PROOF: We may assume that $T = f[S]$. Then f has an inverse f^{-1}. A set $A \subseteq S$ satisfies $\operatorname{diam} f[A] = r \operatorname{diam} A$. Therefore $(\operatorname{diam} f[A])^s = r^s (\operatorname{diam} A)^s$. By the Method I theorem (applied twice), $\overline{\mathcal{H}}^s_{r\varepsilon}(f[F]) = r^s \overline{\mathcal{H}}^s_\varepsilon(F)$. Therefore $\mathcal{H}^s(f[F]) = r^s \mathcal{H}^s(F)$ and $\dim f[F] = \dim F$. ☺

(6.1.9) EXERCISE. *Suppose $f \colon S \to T$ is a function. Let $A \subseteq S$ be a Borel set. Prove or disprove: (1) If f has bounded increase, then*

$$\dim f[A] \leq \dim A.$$

(2) If f has bounded decrease, then

$$\dim f[A] \geq \dim A.$$

(6.1.10) EXERCISE. *Suppose S is a metric space and $\dim S < \infty$. Does it follow that S is separable?*

6.2. EXAMPLES

According to Mandelbrot's tentative definition, a **fractal** is a set A with ind $A <$ dim A. In order for this to be useful, we will have to be able to compute the dimensions involved. In some cases this is not easy to do.

In this section, we will do a few examples directly from the definition. In Section 6.3 we will discuss self-similar sets in general.

Here is our first official example of a fractal. Recall that we evaluated the small inductive dimension ind $\{0,1\}^{(\omega)} = 0$ in Theorem 3.1.9.

(6.2.1) PROPOSITION. *Let $E = \{0,1\}$ be a two-letter alphabet, let $E^{(\omega)}$ be the space of all infinite strings using E, and let $\rho_{1/2}$ be the metric for $E^{(\omega)}$. The Hausdorff dimension of $\left(E^{(\omega)}, \rho_{1/2}\right)$ is 1.*

PROOF: Recall the measure $\mathcal{M}_{1/2}$ on $E^{(\omega)}$ (Proposition 5.5.1). I will show that $\mathcal{H}^1 = \mathcal{M}_{1/2}$. Since $\mathcal{M}_{1/2}(E^{(\omega)}) = 1$, this will show that dim $E^{(\omega)} = 1$. To prove that $\mathcal{H}^1 = \mathcal{M}_{1/2}$, I will use two applications of the Method I theorem.

If a set $A \subseteq E^{(\omega)}$ has positive diameter, then (Proposition 2.5.7) there is a string $\alpha \in E^{(*)}$ with $A \subseteq [\alpha]$ and diam $A =$ diam $[\alpha]$. So $\mathcal{M}_{1/2}(A) \leq \mathcal{M}_{1/2}([\alpha]) =$ diam $[\alpha] =$ diam A. But $\overline{\mathcal{H}}_\varepsilon^1$ is the largest outer measure with $\overline{\mathcal{H}}_\varepsilon^1(A) \leq$ diam A for all sets A of diameter $\leq \varepsilon$. So $\overline{\mathcal{M}}_{1/2} \leq \overline{\mathcal{H}}_\varepsilon^1$. This is true for all $\varepsilon > 0$, so $\overline{\mathcal{M}}_{1/2} \leq \overline{\mathcal{H}}^1$.

On the other hand, let $\alpha \in E^{(*)}$ be a finite string, and $\varepsilon > 0$. There is n so large that $2^{-n} < \varepsilon$ and $n \geq |\alpha|$, the length of α. Then the basic open set $[\alpha]$ is the disjoint union of all sets $[\beta]$, where $\beta \geq \alpha$ and $|\beta| = n$. There are $2^{n-|\alpha|}$ of these sets. Then

$$\overline{\mathcal{H}}_\varepsilon^1([\alpha]) \leq \sum_{\substack{\beta \geq \alpha \\ |\beta| = n}} \text{diam}\,[\beta] = \sum_{\substack{\beta \geq \alpha \\ |\beta| = n}} 2^{-n} = 2^{-|\alpha|}.$$

But $\overline{\mathcal{M}}_{1/2}$ is the largest outer measure with $\overline{\mathcal{M}}_{1/2}([\alpha]) \leq 2^{-|\alpha|}$ for all $\alpha \in E^{(*)}$. So $\overline{\mathcal{H}}_\varepsilon^1 \leq \overline{\mathcal{M}}_{1/2}$, and thus $\overline{\mathcal{H}}^1 \leq \overline{\mathcal{M}}_{1/2}$.

Therefore $\overline{\mathcal{H}}^1 = \overline{\mathcal{M}}_{1/2}$. The measurable sets in both cases are given by the criterion of Carathéodory, so $\mathcal{H}^1 = \mathcal{M}_{1/2}$. ☺

The computation used in the preceding proof will be used again in the future. It is written here for reference. The details are left to the reader.

(6.2.2) THEOREM. *Let (V, E, i, t) be a directed multigraph. Suppose non-negative numbers w_α satisfy*

$$w_\alpha = \sum_{i(e)=t(\alpha)} w_{\alpha e}$$

for $\alpha \in E_v^{(*)}$. Let $\mathcal{M}$ be the method I measure with $\mathcal{M}([\alpha]) = w_\alpha$. If ρ is a metric on $E_v^{(*)}$ and $s > 0$ satisfies $\mathcal{M}([\alpha]) = (\text{diam}\,[\alpha])^s$ for all $\alpha \in E_v^{(*)}$, then $\mathcal{M}(B) = \mathcal{H}^s(B)$ for all Borel sets $B \subseteq E_v^{(*)}$.

Next is our first official example of a non-fractal. We proved ind $\mathbf{R} = 1$ in Theorem 3.1.5.

(6.2.3) PROPOSITION. *The Hausdorff dimension of the line* $\mathbf{R}$ *is 1.*

PROOF: By Theorem 6.1.4, we have $\mathcal{H}^1([0,1]) = \mathcal{L}([0,1]) = 1$. Therefore $\dim[0,1] = 1$. Now $[0,1] \subseteq \mathbf{R}$, so $\dim \mathbf{R} \geq \dim[0,1] = 1$. If $s > 1$, then $\mathcal{H}^s([0,1]) = 0$. The intervals $[n, n+1]$ are isometric to $[0,1]$, so it follows that $\mathcal{H}^s([n, n+1]) = 0$. Therefore

$$\mathcal{H}^s(\mathbf{R}) \leq \sum_{n=-\infty}^{\infty} \mathcal{H}^s([n, n+1]) = 0.$$

This means that $\dim \mathbf{R} \leq s$. But this is true for any $s > 1$, so $\dim \mathbf{R} \leq 1$. Therefore we have seen that $\dim \mathbf{R} = 1$. ☺

Lebesgue measure vs. Hausdorff measure. Since the Lebesgue measure was useful in computing $\dim \mathbf{R}$, it is easy to guess that $\mathcal{L}^2$ is useful in computing $\dim \mathbf{R}^2$.

(6.2.4) PROPOSITION. *The Hausdorff dimension of two-dimensional Euclidean space* $\mathbf{R}^2$ *is 2.*

PROOF: Consider the unit square $Q = [0,1] \times [0,1]$. It is covered by n^2 small squares with side $1/n$, so if $\varepsilon \geq \sqrt{2}/n$, we have $\overline{\mathcal{H}}_\varepsilon^2(Q) \leq n^2(\sqrt{2}/n)^2 = 2$. Therefore $\mathcal{H}^2(Q) \leq 2$, so $\dim Q \leq 2$.

On the other hand, if $\mathcal{A}$ is any cover of Q by closed sets, then (since any set A of diameter r is contained in a square Q_A with side $\leq r$),

$$\sum_{A \in \mathcal{A}} (\text{diam}\,A)^2 \geq \sum_{A \in \mathcal{A}} \mathcal{L}^2(Q_A)$$

$$\geq \mathcal{L}^2 \left(\bigcup_{A \in \mathcal{A}} Q_A \right)$$

$$\geq \mathcal{L}^2(Q) = 1.$$

Therefore $\mathcal{H}^2(Q) \geq 1$, so $\dim Q \geq 2$.

For $\mathbf{R}^2$, since $Q \subseteq \mathbf{R}^2$, we have $\dim \mathbf{R}^2 \geq \dim Q = 2$. If $s > 2$, then $\mathcal{H}^s(Q) = 0$; but $\mathbf{R}^2$ can be covered by a countable collection $\{\, Q_n : n \in \mathbf{N} \,\}$ of squares of side 1, so $\mathcal{H}^s(\mathbf{R}^2) \leq \sum_n \mathcal{H}^s(Q_n) = 0$. This shows that $\dim \mathbf{R}^2 \leq s$. Therefore $\dim \mathbf{R}^2 \leq 2$. ☺

Note that the proof showed $0 < \mathcal{H}^2(Q) < \infty$, where Q is the unit square.

What is the connection between the two measures $\mathcal{L}^2$ and $\mathcal{H}^2$ on $\mathbf{R}^2$? In fact, one of them is just a constant multiple of the other. The proof, however, seems to require a bit more knowledge of two-dimensional geometry than I have generally assumed here.* I will prove only:

---

*See page 187.

(6.2.5) THEOREM. *There exist positive constants a and b such that*

$$a\,\mathcal{L}^2(B) \le \mathcal{H}^2(B) \le b\,\mathcal{L}^2(B)$$

for all Borel sets $B \subseteq \mathbf{R}^2$.

PROOF: Let $Q = [0,1] \times [0,1]$ be the unit square. Let $b = \mathcal{H}^2(Q)$. (We have seen that $b < \infty$.) First, if $B = rQ = [0,r] \times [0,r]$, then $\mathcal{H}^2(B) = r^2\,\mathcal{H}^2(Q) = r^2 b = b\mathcal{L}^2(B)$. Next, the same is true for a translate of such a square. Then, if B is a rectangle with edges parallel to the coordinate axes, it can be approximated inside and outside by little squares, so we will have $\mathcal{H}^2(B) = b\mathcal{L}^2(B)$ in that case. From this we may deduce by the Method I theorem: For any Borel set B, we have $\mathcal{H}^2(B) \le b\mathcal{L}^2(B)$.

Let $a = 1/\mathcal{L}^2(B_1(0))\ [\,= 1/\pi]$. If a set $A \subseteq \mathbf{R}^2$ has diameter r, then it is contained in a ball $B_r(x)$ of radius r, and $\mathcal{L}^2(A) \le \mathcal{L}^2(B_r(x)) = r^2\,\mathcal{L}^2(B_1(0)) = r^2/a$. From this we may deduce by the Method I theorem: For any Borel set B, we have $a\,\mathcal{L}^2(B) \le \mathcal{H}^2(B)$. ☺

Let d be a positive integer. The same method may be used to prove that there exist positive constants a_d and b_d such that

$$a_d\mathcal{L}^d(B) \le \mathcal{H}^d(B) \le b_d\mathcal{L}^d(B)$$

for all Borel sets $B \subseteq \mathbf{R}^d$. The elementary consequences of this are left to the reader:

(6.2.6) EXERCISE. *If $B \subseteq \mathbf{R}^d$, then dim $B \le d$. If B contains an open ball, then dim $B = d$.*

Arc length. Let $f\colon [0,1] \to S$ be a continuous curve in S. The **arc length** of the curve is

$$\sup \sum_{i=1}^{n} \rho\big(f(x_{i-1}), f(x_i)\big),$$

where the supremum is over all finite subdivisions

$$0 = x_0 < x_1 < \cdots < x_n = 1$$

of the interval $[0,1]$. If the arc length is finite, then we say that the curve is **rectifiable**.

(6.2.7) THEOREM. *Let $f\colon [0,1] \to S$ be a continuous curve, let l be its arc length, and write $C = f\big[[0,1]\big]$.*

  (a) $l \ge \mathcal{H}^1(C)$;
  (b) *If f is one-to-one, then $l = \mathcal{H}^1(C)$.*

PROOF: (a) Let $\varepsilon > 0$. Now f is uniformly continuous (Theorem 2.2.21), so there is $\delta > 0$ such that $\rho(f(x), f(y)) < \varepsilon/2$ whenever $|x - y| < \delta$. Choose a subdivision

$$0 = x_0 < x_1 < \cdots < x_n = 1$$

of $[0, 1]$ with $|x_i - x_{i-1}| < \delta$ for all i. Then the sets

$$A_i = f\Big[[x_{i-1}, x_i]\Big]$$

cover C. (But diam A_i may not be $\rho(f(x_{i-1}), f(x_i))$.) By the compactness of $[x_{i-1}, x_i]$, there exist y_i, z_i with $x_{i-1} \le y_i < z_i \le x_i$ such that diam $A_i = \rho(f(y_i), f(z_i))$. Now we may use the subdivision

$$0 \le y_1 < z_1 \le y_2 < z_2 \le \cdots \le y_n < z_n \le 1$$

to estimate the length. So

$$l \ge \sum_{i=1}^n \rho(f(y_i), f(z_i)) = \sum_{i=1}^n \text{diam } A_i \ge \overline{\mathcal{H}}^1_\varepsilon(C).$$

Now let $\varepsilon \to 0$ to obtain $l \ge \mathcal{H}^1(C)$.

(b) First, I claim that if $0 \le a < b \le 1$, then $\mathcal{H}^1\big(f[[a, b]]\big) \ge \rho(f(a), f(b))$. To see this, consider the function $h\colon f[[a, b]] \to \mathbb{R}$ defined by $h(u) = \rho(f(a), u)$. Now h is continuous, and h has values $h(a) = 0$ and $h(b) = \rho(f(a), f(b))$, so by the intermediate value theorem, applied to the continuous function $h \circ f\colon [a, b] \to \mathbb{R}$, we know that h also has all values between. Now h satisfies $|h(u) - h(v)| \le \rho(u, v)$, so we have

$$\mathcal{H}^1\Big(f[[a, b]]\Big) \ge \mathcal{H}^1\Big(h[f[[a, b]]]\Big)$$
$$\ge \mathcal{H}^1\Big([0, \rho(f(a), f(b))]\Big)$$
$$= \rho(f(a), f(b)).$$

This proves the claim.

Now we apply this inequality. If we have a subdivision

$$0 = x_0 < x_1 < \cdots < x_n = 1$$

of $[0, 1]$, then the sets $f[[x_{i-1}, x_i]]$ are measurable and disjoint, since f is one-to-one. So

$$\sum_{i=1}^n \rho(f(x_{i-1}), f(x_i)) \le \sum_{i=1}^n \mathcal{H}^1\Big(f[[x_{i-1}, x_i]]\Big)$$
$$= \mathcal{H}^1\left(\bigcup_{i=1}^n f[[x_{i-1}, x_i]]\right)$$
$$= \mathcal{H}^1\Big(f[[0, 1]]\Big) \le \mathcal{H}^1(C).$$

This is true for all subdivisions, so $l \le \mathcal{H}^1(C)$. ☺

(6.2.8) EXERCISE. *What is the relation between the surface area (of a surface in $\mathbb{R}^3$) and its two-dimensional Hausdorff measure?*

Hausdorff dimension vs. topological dimension. We will see in the next section that it is possible for the Hausdorff dimension dim F to have a non-integer value. But it is not completely unrelated to the topological dimension.

(6.2.9) THEOREM. *Let S be a metric space. Then* ind $S \leq$ dim S.

A complete proof of this result can be found, for example, in [**30**, p. 104]. (The proof uses Lebesgue integration, which we have avoided in this book.) Here, we will prove it for compact spaces:

(6.2.10) THEOREM. *Let S be a compact metric space. Then* ind $S \leq$ dim S.

PROOF: We know (Theorem 3.4.7) that ind $S =$ Cov S. Let $n =$ Cov S. This means that Cov $S \leq n - 1$ is false. So there exist open sets $U_1, U_2, \cdots, U_{n+1}$ such that $\bigcup_{i=1}^{n+1} U_i = S$, but for any closed sets $F_i \subseteq U_i$ with $\bigcup_{i=1}^{n+1} F_i = S$, we must have $\bigcap_{i=1}^{n+1} F_i \neq \varnothing$.

Define functions on S as follows:

$$d_i(x) = \text{dist}(x, S \setminus U_i), \qquad 1 \leq i \leq n+1$$
$$d(x) = d_1(x) + d_2(x) + \cdots + d_{n+1}(x).$$

The functions are continuous; in fact

$$|d_i(x) - d_i(y)| \leq \rho(x, y),$$
$$|d(x) - d(y)| \leq (n+1)\rho(x, y).$$

Since the sets U_i cover S, we have $d(x) > 0$ for all x. So since S is compact, there exist positive constants a, b such that $a \leq d(x) \leq b$ for all $x \in S$. Now define $h: S \to \mathbb{R}^{n+1}$ by

$$h(x) = \left(\frac{d_1(x)}{d(x)}, \frac{d_2(x)}{d(x)}, \cdots, \frac{d_{n+1}(x)}{d(x)} \right).$$

The function h has bounded increase:

$$\left| \frac{d_i(x)}{d(x)} - \frac{d_i(y)}{d(y)} \right| = \frac{|d(x)d_i(y) - d(y)d_i(x)|}{d(x)d(y)}$$
$$\leq \frac{d(x)\,|d_i(y) - d_i(x)| + d_i(x)\,|d(x) - d(y)|}{d(x)d(y)}$$
$$\leq \frac{b(n+2)}{a^2}\rho(x, y),$$

and therefore

$$|h(x) - h(y)| \leq \frac{b(n+1)(n+2)}{a^2}\rho(x, y).$$

Now I claim that $h[S]$ includes the simplex

$$T = \left\{ (t_1, t_2, \cdots, t_{n+1}) \in \mathbf{R}^{n+1} : t_i > 0, \sum_{i=1}^{n+1} t_i = 1 \right\}.$$

Given $(t_1, t_2, \cdots, t_{n+1}) \in T$, consider the sets

$$F_i = \left\{ x : \frac{d_i(x)}{d(x)} \geq t_i \right\}.$$

Then F_i is closed, $F_i \subseteq U_i$, and $\bigcup_{i=1}^{n+1} F_i = S$ since $\sum_i d_i(x)/d(x) = 1$. So we know by hypothesis that $\bigcap_{i=1}^{n+1} F_i \neq \varnothing$. That is, there exists a point $x \in S$ with $d_i(x)/d(x) \geq t_i$ for all i. But since $\sum_i d_i(x)/d(x) = 1$ we have $d_i(x)/d(x) = t_i$ for all i. That is, $h(x) = (t_1, t_2, \cdots t_{n+1})$. So $h[S] \supseteq T$.

Now T is isometric to an open set in $\mathbf{R}^n$. By Theorem 6.1.6 and Exercise 6.1.9, we have $\dim S \geq \dim T = n$. ☺

(6.2.11) EXERCISE. *Let S be a metric space. If* $\operatorname{ind} S \geq 1$, *then* $\dim S \geq 1$.

6.3. SELF-SIMILARITY

Self-similarity is one of the easiest ways to produce examples of fractals. This section deals with the question of when the similarity dimension can be used to compute the Hausdorff dimension. When the two coincide we have a desirable situation: the similarity dimension is easy to compute, and the Hausdorff dimension is more generally applicable and has many useful properties.

Let $(r_1, r_2, \cdots, r_n)$ be a contracting ratio list. Let $(f_1, f_2, \cdots, f_n)$ be a corresponding iterated function system of similarities in $\mathbf{R}^d$. Suppose K is the invariant set for the iterated function system. Does it follow that $\dim K = s$? In general, the answer is no. There is an inequality $\dim K \leq s$. But simple examples show that if there is "too much" overlap among the pieces $f_i[K]$, then $\dim K < s$ is possible.

We will consider Moran's open set condition, which is a general condition that insures $\dim K = s$.

Cantor dust. First we will discuss the Hausdorff dimension of the triadic Cantor dust (defined on page 2). The ratio list for this set is $(1/3, 1/3)$. The string model is the set $E^{(\omega)}$ of infinite strings from the alphabet $E = \{\mathbf{0}, \mathbf{1}\}$, together with the metric $\rho_{1/3}$. The two similarities on the model space are the right shifts, which we will call θ_0 and θ_1. They are defined as follows:

$$\theta_0(\sigma) = \mathbf{0}\sigma$$
$$\theta_1(\sigma) = \mathbf{1}\sigma.$$

Thus (θ_0, θ_1) is a realization of the ratio list $(1/3, 1/3)$, with invariant set $E^{(\omega)}$.

(6.3.1) PROPOSITION. *The Hausdorff dimension for* $E^{(\omega)}$ *with metric* $\rho_{1/3}$ *is* $\log 2/\log 3$.

PROOF: Write $s = \log 2/\log 3$. Consider the metric $\rho_{1/3}$ and the measure $\mathcal{M}_{1/2}$. If the length $|\alpha| = k$, then $\mathcal{M}_{1/2}([\alpha]) = 2^{-k} = (3^{-k})^s = (\mathrm{diam}\,[\alpha])^s$. Therefore, by Theorem 6.2.2, $\mathcal{H}^s(E^{(\omega)}) = \mathcal{M}_{1/2}(E^{(\omega)}) = 1$; so $\dim E^{(\omega)} = s = \log 2/\log 3$. ☺

(6.3.2) COROLLARY. *The Cantor dust has Hausdorff dimension* $\log 2/\log 3$.

PROOF: A map with bounded distortion preserves the Hausdorff dimension (Exercise 6.1.9). The model map h from $E^{(\omega)}$ onto the triadic Cantor dust C has bounded distortion (Proposition 2.5.3). ☺

Sierpiński gasket. Next we discuss a slightly more difficult example, the Sierpiński gasket (see page 8).

Let S be the Sierpiński gasket. It is the invariant set for an iterated function system with ratio list $(1/2, 1/2, 1/2)$. Let s [$= \log 3/\log 2$] be the dimension of the ratio list. Let $E = \{\mathbf{L}, \mathbf{U}, \mathbf{R}\}$ be the appropriate three-letter alphabet. We will need a metric and a measure defined on $E^{(\omega)}$.

Let ρ be the metric on $E^{(\omega)}$ for the ratio list $(1/2, 1/2, 1/2)$. That is, ρ is defined so that $\mathrm{diam}\,[\alpha] = 2^{-|\alpha|}$ for all $\alpha \in E^{(*)}$. Then the right shifts realize the ratio list:

$$\rho(\mathbf{L}\sigma, \mathbf{L}\tau) = \frac{1}{2}\,\rho(\sigma, \tau),$$

$$\rho(\mathbf{U}\sigma, \mathbf{U}\tau) = \frac{1}{2}\,\rho(\sigma, \tau),$$

$$\rho(\mathbf{R}\sigma, \mathbf{R}\tau) = \frac{1}{2}\,\rho(\sigma, \tau).$$

The measure $\mathcal{M}$ is specified by $\mathcal{M}([\alpha]) = 3^{-|\alpha|}$. Each node in $E^{(*)}$ has exactly 3 children, so these numbers satisfy the required additivity (Theorem 5.5.4). The fact to notice is this:

$$\mathcal{M}([\alpha]) = \left(\mathrm{diam}\,[\alpha]\right)^s$$

for all $\alpha \in E^{(*)}$, where $s = \log 3/\log 2$.

The following is proved in the same way as Proposition 6.3.1, using this metric ρ and this measure $\mathcal{M}$:

(6.3.3) EXERCISE. *The Hausdorff dimension of the string space* $E^{(\omega)}$ *defined above is* $s = \log 3/\log 2$.

(6.3.4) EXERCISE. *Let s be any positive real number. There is a metric space S with* $\dim S = s$.

The dimension calculation for the string model will be used to help with the dimension calculation of the Sierpiński gasket S itself.

Let $h: E^{(\omega)} \to \mathbf{R}^2$ be the model map that sends $E^{(\omega)}$ onto the gasket S. If the iterated function system in $\mathbf{R}^2$ is $(f_\mathsf{L}, f_\mathsf{U}, f_\mathsf{R})$, then

$$h(\mathsf{L}\sigma) = f_\mathsf{L}\big(h(\sigma)\big),$$
$$h(\mathsf{U}\sigma) = f_\mathsf{U}\big(h(\sigma)\big),$$
$$h(\mathsf{R}\sigma) = f_\mathsf{R}\big(h(\sigma)\big).$$

To simplify the notation, we will write $\mathsf{L}(x)$ in place of $f_\mathsf{L}(x)$, and similarly for the other two letters, and write $\alpha(x)$ for a finite string α.

(6.3.5) PROPOSITION. *The Sierpiński gasket has Hausdorff dimension at most* $\log 3/\log 2$.

PROOF: The model map h has bounded increase (Exercise 4.2.1). By Exercise 6.1.9, we have dim $S \le$ dim $E^{(\omega)}$. ☺

The model map for the Sierpiński gasket does not have bounded decrease. In fact, it is not even one-to-one. (This is the answer to Exercise 4.2.2.) So we will need a bit more effort to prove the lower bound. Pay attention to the ingredients of the proof, since they will be used again for the general case.

(6.3.6) PROPOSITION. *The Sierpiński gasket S has Hausdorff dimension equal to the similarity dimension* $\log 3/\log 2$.

PROOF: Let V be the interior of the first triangle S_0 approximating the Sierpiński gasket S. Then $\mathcal{L}^2(V) = \sqrt{3}/4$, and if $|\alpha| = |\beta|$, $\alpha \ne \beta$, then $\alpha[V] \cap \beta[V] = \varnothing$. Also, $h\big[[\alpha]\big] = \overline{\alpha[V]}$. The set S_k approximating S is the union

$$\bigcup_{\alpha \in E^{(k)}} \overline{\alpha[V]}.$$

Given a set $A \subseteq S$, let k be the positive integer satisfying

$$2^{-k} < \operatorname{diam} A \le 2^{-k+1}.$$

Let

$$T = \left\{ \alpha \in E^{(k)} : \overline{\alpha[V]} \cap A \ne \varnothing \right\}.$$

I claim that T has at most 100 elements. Let m be the number of elements in T. A set $\alpha[V]$ is the image of V under a similarity with ratio 2^{-k}, so it has area

$$\mathcal{L}^2\big(\alpha[V]\big) = 4^{-k}\frac{\sqrt{3}}{4}.$$

The sets $\alpha[V]$ with $\alpha \in T$ are all disjoint. If x is a point of A, then all of the elements of all of the sets $\alpha[V]$ with $\alpha \in T$ are within distance diam $A + 2^{-k} \le$

$3 \cdot 2^{-k}$ of x. So m disjoint sets of area $4^{-k}\sqrt{3}/4$ are contained in the ball with center x and radius $3 \cdot 2^{-k}$. Therefore

$$m4^{-k}\frac{\sqrt{3}}{4} \leq \pi(3 \cdot 2^{-k})^2.$$

Solving for m, we get $m \leq 36\pi/\sqrt{3}$, which is smaller than 100.

Next I claim $\mathcal{M}(h^{-1}[A]) \leq 100(\operatorname{diam} A)^s$ for all Borel sets $A \subseteq S$. Given A, let k and T be as above. Then $A \subseteq \bigcup_{\alpha \in T} \overline{\alpha[V]}$, so $h^{-1}[A] \subseteq \bigcup_{\alpha \in T} [\alpha]$. Therefore

$$\mathcal{M}(h^{-1}[A]) \leq \sum_{\alpha \in T} \mathcal{M}([\alpha])$$
$$\leq 100 \cdot 3^{-k} = 100(2^{-k})^s$$
$$\leq 100(\operatorname{diam} A)^s.$$

By the Method I theorem, $\mathcal{M}(h^{-1}[A]) \leq 100\mathcal{H}^s(A)$ for all Borel sets A. So $1 \leq 100\mathcal{H}^s(S)$, and therefore $\dim S \geq s$. ☺

General ratio lists. Let $(r_1, r_2, \cdots, r_n)$ be a contracting ratio list with dimension s. Let $(f_1, f_2, \cdots, f_n)$ be a corresponding iterated function system of similarities in $\mathbb{R}^d$. Suppose K is the invariant set for the iterated function system.

In general it is not true that $\dim K = s$. For example, consider the iterated function system (f_L, f_U, f_R) for the Sierpiński gasket, realizing the ratio list $(1/2, 1/2, 1/2)$. Now the iterated function system (f_L, f_L, f_U, f_R) has the same invariant set, of course, but it realizes the longer ratio list $(1/2, 1/2, 1/2, 1/2)$. The Hausdorff dimension of the invariant set K is $\log 3/\log 2$, but the similarity dimension of the iterated function system is 2.

Of course, the problem is that the first two images $f_L[K]$ and $f_L[K]$ overlap too much. Now we might require that the images do not overlap at all, as in the Cantor dust. But that would rule out many of the most interesting examples, such as the Sierpiński gasket itself, where the overlap sets like $f_L[K] \cap f_U[K]$ are nonempty.

We do have an inequality between the Hausdorff dimension and the similarity dimension. This is the "upper bound" calculation $\dim K \leq s$.

Upper bound. Now begin with a contracting ratio list $(r_1, r_2, \cdots, r_n)$, with $n > 1$. Then the dimension s associated with it is the unique positive number s satisfying

$$\sum_{i=1}^{n} r_i^s = 1.$$

Let E be an n-letter alphabet, and let $E^{(\omega)}$ be the string model. The metric ρ on E is defined so that the right shifts realize the given ratio list. We define $r(\alpha)$ recursively, starting with the empty string Λ, by:

$$r(\Lambda) = 1,$$
$$r(\alpha e) = r(\alpha) \cdot r_e,$$

then define ρ so that $\operatorname{diam}[\alpha] = r(\alpha)$.

We will also need a measure defined to fit the ratio list. The basis is the equation defining the dimension s:

$$\sum_{i=1}^{n} r_i^s = 1.$$

It follows from this that

$$\sum_{i=1}^{n} \left(r(\alpha)r_i\right)^s = r(\alpha)^s.$$

That is, the expression $r(\alpha)^s$ satisfies the additivity condition for a metric outer measure. The measure $\mathcal{M}$ in question will be defined on the string space $E^{(\omega)}$, and satisfy $\mathcal{M}([\alpha]) = r(\alpha)^s$ for all α. Of course, it is no coincidence that s was chosen so that $\mathcal{M}([\alpha]) = (\operatorname{diam}[\alpha])^s$.

By Theorem 6.2.2 we have $\mathcal{M}(B) = \mathcal{H}^s(B)$ for all Borel sets $B \subseteq E^{(\omega)}$. This proves:

(6.3.7) THEOREM. *The Hausdorff dimension of the string model $E^{(\omega)}$ is equal to the similarity dimension s.*

(6.3.8) THEOREM. *Let K be the invariant set of a realization in a complete metric space S of a contracting ratio list with dimension s. Then* $\dim K \leq s$.

PROOF: The model map $h\colon E^{(*)} \to S$ has bounded increase (Exercise 4.2.4). ☺

(6.3.9) EXERCISE. *Let $(r_1, r_2, \cdots, r_n)$ be a contracting ratio list with dimension s. Let $(f_1, f_2, \cdots, f_n)$ be an iterated function system consisting not of similarities, but merely satisfying*

$$\rho\big(f_i(x), f_i(y)\big) \leq r_i\, \rho(x, y).$$

Then the invariant set K has Hausdorff dimension $\leq s$.

Lower bound. The iterated function system $(f_1, f_2, \cdots, f_n)$ satisfies **Moran's open set condition** iff there exists a nonempty open set U for which we have $f_i[U] \cap f_j[U] = \varnothing$ for $i \neq j$ and $U \supseteq f_i[U]$ for all i.

For example, consider the Cantor dust. The similarities are

$$f_0(x) = \frac{x}{3},$$
$$f_1(x) = \frac{x+2}{3}.$$

The open set $U = (0,1)$ satisfies the open set condition: the two images are $(0, 1/3)$ and $(2/3, 1)$, which are disjoint and contained in U.

Or, consider the Sierpiński gasket (Figure 1.2.1). The interior† U of the large triangle S_0 satisfies the open set condition. The three images are three small triangles, contained in U, and disjoint.

For a third example, consider the Koch curve (Figure 1.5.1). The interior of the triangle L_0 satisfies the open set condition.

The fourth example to consider is Heighway's dragon. This time the open set condition is not quite as trivial. The interior U of Heighway's dragon itself will serve. The fact that the two images are contained in U is a consequence of the fact that Heighway's dragon itself is the invariant set of the iterated function system. The fact that the two images are disjoint is a consequence of the fact that the approximating polygon never crosses itself (Proposition 1.5.7). The verification is left to the reader:

(6.3.10) EXERCISE. *The interior of Heighway's dragon satisfies the open set condition for the iterated function system realizing Heighway's dragon.*

In one case, the open set condition is easily verified:

(6.3.11) EXERCISE. *If the invariant set K satisfies $f_i[K] \cap f_j[K] = \varnothing$ for $i \neq j$, then Moran's open set condition is satisfied.*

The proof of the lower bound will proceed following the same technique as Proposition 6.3.6. The area is replaced by the d-dimensional volume, namely $\mathcal{L}^d$. You may find it instructive to compare this argument with the proof of the special case in Proposition 6.3.6.

Let E be an alphabet with n letters. Write the ratio list as $(r_e)_{e \in E}$ and the iterated function system as $(f_e)_{e \in E}$. To simplify the notation, we will write $e(x)$ in place of $f_e(x)$, and similarly $\alpha(x)$ for a finite string α. With this notation, the model map $h: E^{(\omega)} \to \mathbb{R}^d$ satisfies $h(e\alpha) = e(h(\alpha))$ for $\alpha \in E^{(*)}$ and $e \in E$.

The open set condition implies that $\alpha[U] \cap \beta[U] = \varnothing$ for two strings $\alpha, \beta \in E^{(*)}$ unless one is an initial segment of the other. If α is a string with length $k \geq 1$, we will write α^- for the parent of α; that is: $\alpha^- = \alpha \upharpoonright (k - 1)$.

(6.3.12) THEOREM. *Let $(r_e)_{e \in E}$ be a contracting ratio list. Let s be its dimension, and let $(f_e)_{e \in E}$ be a realization in $\mathbb{R}^d$. Let K be the invariant set. If Moran's open set condition is satisfied, then $\dim K = s$.*

PROOF: First I claim that there is a constant $c > 0$ such that: if $A \subseteq K$, then the set

$$T = \left\{ \alpha \in E^{(*)} : \overline{\alpha[U]} \cap A \neq \varnothing, \ \operatorname{diam} \alpha[U] < \operatorname{diam} A \leq \operatorname{diam} \alpha^-[U] \right\}$$

has at most c elements.

As α ranges over T, the sets $\alpha[U]$ are disjoint, since no such α is an initial segment of another. The map f_α is a similarity with ratio equal to $\operatorname{diam}[\alpha]$, so

---

†The interior of a set consists of all the interior points of the set.

if w is the diameter of U, then $w \operatorname{diam}[\alpha]$ is the diameter of $\alpha[U]$. If $r = \min r_e$, then

$$\operatorname{diam} \alpha[U] = w \operatorname{diam}[\alpha] \geq wr \operatorname{diam}[\alpha^-]$$
$$= r \operatorname{diam} \alpha^-[U] \geq r \operatorname{diam} A.$$

If $p = \mathcal{L}^d(U)$ is the volume of U, then the volume of $\alpha[U]$ is

$$\mathcal{L}^d(\alpha[U]) = p \cdot \left(\frac{\operatorname{diam} \alpha[U]}{\operatorname{diam} U}\right)^d \geq \frac{pr^d}{w^d}(\operatorname{diam} A)^d$$

If x is a point of A, then every point of every set $\alpha[U]$ for $\alpha \in T$ is within distance $\operatorname{diam} A + \operatorname{diam} \alpha[U] \leq 2 \operatorname{diam} A$ of x. If m is the number of elements of T, then we have m disjoint sets $\alpha[U]$, all with volume at least $(pr^d/w^d)(\operatorname{diam} A)^d$, contained within a ball of radius $2 \operatorname{diam} A$. So if $t = \mathcal{L}^d(B_1(0))$ is the volume of the unit ball, we have

$$\frac{mpr^d}{w^d}(\operatorname{diam} A)^d \leq t(2 \operatorname{diam} A)^d.$$

Solving for m yields

$$m \leq \frac{tw^d 2^d}{pr^d}.$$

Summary: We may use the constant $c = tw^d 2^d / pr^d$, where $r = \min r_e$, t is the volume of the unit ball, p is the volume of U, and w is the diameter of U.

Next I claim there is a positive constant b so that for any Borel set $A \subseteq K$, we have

$$\mathcal{M}(h^{-1}[A]) \leq b(\operatorname{diam} A)^s.$$

Given A, let T be as above. So $A \subseteq \bigcup_{\alpha \in T} \alpha[U]$, and $h^{-1}[A] \subseteq \bigcup_{\alpha \in T}[\alpha]$. If $\alpha \in T$, then $\mathcal{M}([\alpha]) = (\operatorname{diam}[\alpha])^s = ((1/w) \operatorname{diam} \alpha[U])^s \leq (1/w^s)(\operatorname{diam} A)^s$. Therefore

$$\mathcal{M}(h^{-1}[A]) \leq \sum_{\alpha \in T} \mathcal{M}([\alpha])$$
$$\leq c(1/w^s)(\operatorname{diam} A)^s.$$

So $b = c/w^s$ will work.

Therefore, by the Method I theorem, $1 = \mathcal{M}(h^{-1}[K]) \leq b\mathcal{H}^s(K)$, so we have $\dim K \geq s$. ☺

The proof given above clearly used the properties of Lebesgue measure in $\mathbb{R}^d$. What happens in other metric spaces? Readers who know about some exotic metric spaces may like to attempt this:

(6.3.13) EXERCISE. *Let S be a complete metric space (other than $\mathbf{R}^d$). Let $(f_1, f_2, \cdots, f_n)$ be a realization in S of a contracting ratio list $(r_1, r_2, \cdots, r_n)$ with dimension s. Let K be the invariant set. Suppose Moran's open set condition is satisfied. Does it follow that $\dim K = s$?*

(6.3.14) EXERCISE. *Let $(r_1, r_2, \cdots, r_n)$ be a contracting ratio list with dimension s. Let $(f_1, f_2, \cdots, f_n)$ be an iterated function system consisting not of similarities, but of maps $f_i : \mathbf{R}^d \to \mathbf{R}^d$ satisfying*

$$\rho\big(f_i(x), f_i(y)\big) \geq r_i\,\rho(x, y).$$

Suppose the open set conditon holds, and suppose there is an invariant set K. Does it follow that $\dim K \geq s$?

Let E be a finite alphabet with at least two letters, let $(r_e)_{e \in E}$ be a contracting ratio list, and let $(f_e)_{e \in E}$ be an iterated function system realizing the ratio list in $\mathbf{R}^d$. For finite strings $\alpha \in E^{(*)}$, define maps f_α as usual: f_Λ is the identity and $f_{\alpha e} = f_\alpha \circ f_e$. We say that the iterated function system (f_e) satisfies **Marion's open set condition** iff there is a nonempty open set U such that

$$f_\alpha[U] \cap f_\beta[U] = \varnothing$$

unless one of α, β is an initial segment of the other.

(6.3.15) EXERCISE. *(1) Can Moran's open set condition be replaced by Marion's open set condition in Theorem 6.3.12? (2) Is there an example of an iterated function system satisfying Marion's open set condition but not Moran's open set condition?*

Examples. This is another dragon curve (Figure 6.3.16, Plate 10).

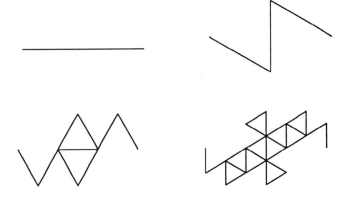

Figure 6.3.16. Terdragon.

```
; terdragon
make "shrink 1/sqrt 3
to ter :depth :size
    if :depth=0 [forward :size stop]
    right 30
    ter :depth-1 :size*:shrink
    left 120
    ter :depth-1 :size*:shrink
    right 120
    ter :depth-1 :size*:shrink
    left 30
end
```

This is a space-filling curve. (Six copies of the terdragon exactly fit around a point; Plate 11.) It is not a fractal. But what about its boundary? Is it a "fractile"?

(6.3.17) EXERCISE. *Compute the Hausdorff dimension for the boundary of the terdragon.*

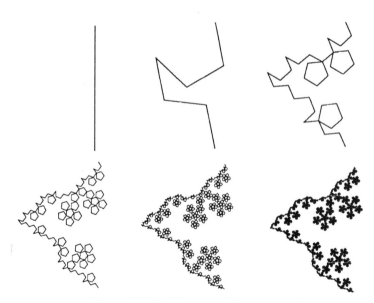

Figure 6.3.18. Pentadendrite construction.

A minor variant of the dragon that constructs the McWorter pentigree (page 24) is shown in Figure 6.3.18. Five copies of the limit set fit together to form a certain "dendrite". This will be called the **pentadendrite**. (Figure 6.3.19).

Figure 6.3.19. Complete pentadendrite.

(6.3.20) EXERCISE. *Compute the topological dimension and Hausdorff dimension of the pentadendrite.*

The set of "fractions" for the number system with base $-1 + i$ and digit set $\{0, 1\}$ is the twindragon (Figure 1.6.7).

(6.3.21) EXERCISE. *Compute the Hausdorff dimension of the boundary of the twindragon.*

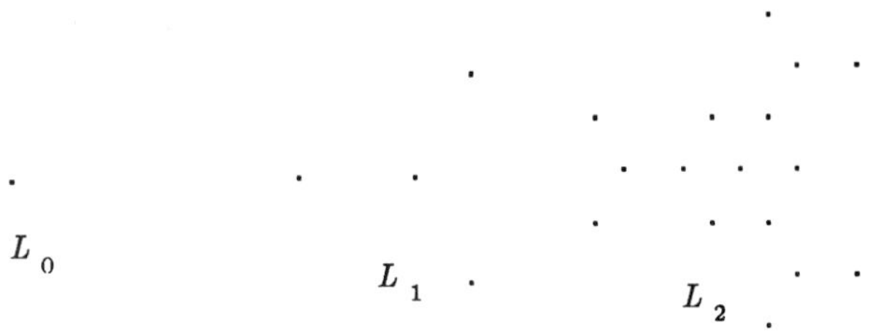

L_0

L_1

L_2

Figure 6.3.22. Eisenstein fractions.

Let us consider next the set F of "fractions" for the Eisenstein number system (Figure 1.6.11). The base is $b = -2$, and the digit set consists of 0, 1, $A = \omega$, and $B = \omega^2$. The set F may be done in the same way as the twindragon (page 30). But let us proceed in a more direct way. The first set L_0 is just the point 0. The next set L_1 consists of the four points $(.0)_{-2}$, $(.1)_{-2}$, $(.A)_{-2}$, and $(.B)_{-2}$. The set L_2 contains 16 points, all that can be represented with two digits in this system. The illustrations in Figure 6.3.22 show these approximations. The set F obtained this time has (of course) similarity dimension 2.

(6.3.23) EXERCISE. *Find the Hausdorff dimension of the boundary of the set F of fractions for this Eisenstein number system.*

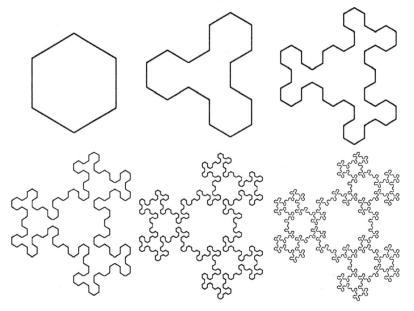

Figure 6.3.24. Boundary.

The construction outlined above is a lot like the construction on page 25 of the second form of McWorter's pentigree. The **pentigree outline** is what we get if we fill in all the "lakes", and take the boundary of the result (Figure 6.3.25).

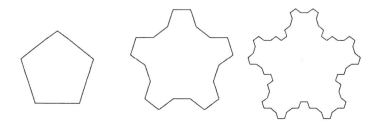

Figure 6.3.25. Pentigree outline.

(6.3.26) EXERCISE. *What is the fractal dimension of the pentigree outline?*

(6.3.27) EXERCISE. *Compute the Hausdorff dimension for the limit of the sets constructed by the following program (page 193). Warning: it is not self-similar.*

```
to Schmidt :depth :size
    if :depth=0 [stop]
    repeat 3 [forward :size Schmidt :depth-1 :size/2 right 120]
end
```

Another example of the same kind is the "I" fractal, page 194.

6.4. GRAPH SELF-SIMILARITY

Next we consider evaluation of the Hausdorff dimension connected with graph self-similar sets.

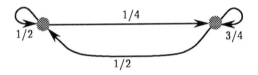

Figure 6.4.1. Graph for the two-part dust.

The two-part dust. We begin with a simple example, the **two-part dust**. It has been "rigged" so that the calculation of the dimension is easier than the general case. The Mauldin-Williams graph is as illustrated in Figure 6.4.1. Here is a description of the realization in $\mathbb{R}^2$ that will be considered. The map **a** has ratio $1/2$, fixed point $(0,0)$, and rotation 30 degrees counterclockwise. The map **b** has ratio $1/4$, fixed point $(1,0)$, and rotation 60 degrees clockwise. The map **c** has ratio $1/2$, fixed point $(0,0)$, and rotation 90 degrees counterclockwise. The map **d** has ratio $3/4$, fixed point $(1,0)$, and rotation 120 degrees clockwise.

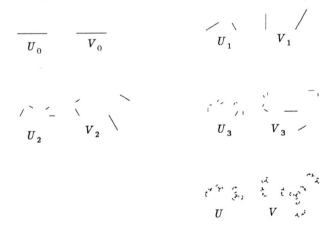

Figure 6.4.2. Two-part dust.

As we know, there is a unique pair of nonempty compact sets $U, V \subseteq \mathbb{R}^2$ satisfying

$$U = \mathbf{a}[U] \cup \mathbf{b}[V]$$
$$V = \mathbf{c}[U] \cup \mathbf{d}[V].$$

This pair of sets is the **two-part dust**. A sequence of approximations is pictured in Figure 6.4.2. They converge in the Hausdorff metric. We may start with any two nonempty compact sets U_0 and V_0 in $\mathbb{R}^2$. In this case, both have been chosen as the line segment from the point $(0,0)$ to the point $(1,0)$. Then further approximations are defined recursively:

$$U_{n+1} = \mathbf{a}[U_n] \cup \mathbf{b}[V_n],$$
$$V_{n+1} = \mathbf{c}[U_n] \cup \mathbf{d}[V_n].$$

The sequence (U_n) converges in the Hausdorff metric to a nonempty compact set U, and the sequence (V_n) converges in the Hausdorff metric to a nonempty compact set V. This pair of sets is the required invariant list.

Here is the Logo program for the pictures.

```
; two-part dust
to U :depth :size
    if :depth=0 [forward :size stop]
    left 30
    U :depth-1 :size/2
    penup
        back :size/2
        right 30
        forward :size
        right 60
        back :size/4
    pendown
    V :depth-1 :size/4 left 60
end
to V :depth :size
    if :depth=0 [forward :size stop]
    left 90
    U :depth-1 :size/2
    penup
        back :size/2
        right 90
        forward :size
        right 120
        back :size*(0.75)
    pendown
    V :depth-1 :size*(0.75) left 120
end
```

We are interested in computing the Hausdorff dimensions of the sets U and V. Since each of the sets is similar to a subset of the other, their dimensions must be the same. As usual, we begin by computing the Hausdorff dimension of the spaces of paths corresponding to the Mauldin-Williams graph.

We will need to use the **Perron numbers** of the graph. In this case, the Perron numbers (one for each node) are $q_U = 1/3$ and $q_V = 2/3$. The important facts about these numbers are: (1) they are positive; and (2) they satisfy equations

(6.4.3)
$$q_U = r(\mathbf{a})q_U + r(\mathbf{b})q_V,$$
$$q_V = r(\mathbf{c})q_U + r(\mathbf{d})q_V.$$

By Proposition 4.3.15, the graph has* dimension 1. I want to show that U and V have Hausdorff dimension 1.

We will use the Perron numbers to assign diameters to the nodes of the path forest. Two ultrametrics ρ, one on each of the two path spaces $E_U^{(\omega)}$, $E_V^{(\omega)}$, will be defined so that the diameters of the basic open sets $[\alpha]$ are as follows: Begin with $\text{diam}[\Lambda_U] = q_U$ and $\text{diam}[\Lambda_V] = q_V$. If α is a path and e is an edge with $t(e) = i(\alpha)$, then the diameter for the string $e\alpha$ is $\text{diam}([e\alpha]) = r(e) \cdot \text{diam}([\alpha])$.

Next, we will use the same numbers to define measures.

There is a metric measure $\mathcal{M}$ on each of the path spaces $E_v^{(\omega)}$ such that $\mathcal{M}([\alpha]) = \text{diam}([\alpha])$ for all finite paths α. The additivity condition is true by equations (6.4.3).

(6.4.4) PROPOSITION. *The path spaces $E_U^{(\omega)}$, $E_V^{(\omega)}$ both have Hausdorff dimension* 1.

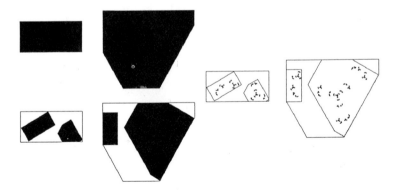

Figure 6.4.5. Open set condition.

---

*Solution to Exercise 4.3.13.

PROOF: The maximum ratio $r = 3/4$ is less than 1. So $r^k \to 0$ as $k \to \infty$. By Theorem 6.2.2, we have $\mathcal{M}(B) = \mathcal{H}^1(B)$ for all Borel sets B. This shows that $\dim E_U^{(\omega)} = 1$ and $\dim E_V^{(\omega)} = 1$. ☺

(Therefore, we say that U and V have [graph] similarity dimension 1.)

But we are more interested in the Hausdorff dimensions of U and V. We will need an "open set condition". A little experimentation with a graphics program reveals that this may be satisfied by the two sets pictured in Figure 6.4.5. The two sets are a rectangle and an irregular hexagon. (The images of these sets under the maps are appropriately disjoint and contained in the appropriate sets.) The dimension is now easy to check.

(6.4.6) EXERCISE. *Let U and V be the two parts of the two part dust. Show that the model maps*

$$h_U : E_U^{(\omega)} \to \mathbf{R}^2$$
$$h_V : E_V^{(\omega)} \to \mathbf{R}^2$$

have bounded distortion.

As consequences of this we have $\dim U = \dim E_U^{(\omega)} = 1$ and $\dim V = \dim E_V^{(\omega)} = 1$.

Perron numbers. To compute the Hausdorff dimension for the other examples with graph self-similarity, we need only find the proper sort of "Perron numbers" in those cases. (It will not be quite as simple as the rigged example above if the dimension is not 1.)

Consider a Mauldin-Williams graph (V, E, i, t, r). We will consider only the case when the graph is strongly connected (page 74). This will mean that when the invariant set list is found, each of the sets will be similar to a subset of each of the others. So they will all have the same Hausdorff dimension.

We are interested in assigning metrics to the spaces $E_v^{(\omega)}$ of strings. (There is one space for each tree in the path forest.) The realization consists, as usual, of the right-shifts. For an edge $e \in E_{uv}$, the function θ_e defined by

$$\theta_e(\sigma) = e\sigma$$

maps $E_v^{(\omega)}$ to $E_u^{(\omega)}$. The metrics should be chosen in such a way that θ_e is a similarity with ratio $r(e)$. We are also interested in defining measures (one for each space $E_v^{(\omega)}$) that will make the computation of the Hausdorff dimension easy.

In order to do this, we need the proper Perron numbers. If s is a positive real number, then s-dimensional **Perron numbers** for the graph are positive numbers q_v, one for each vertex $v \in V$, such that

$$q_u^s = \sum_{\substack{v \in V \\ e \in E_{uv}}} r(e)^s \cdot q_v^s$$

for all $u \in V$.

There is exactly one positive number s such that s-dimensional Perron numbers exist. This unique number s will be called the **dimension** of the Mauldin-Williams graph. The existence and uniqueness of the dimension were proved for the case of a 2 node graph in Section 4.4. For the general graph, the proof requires some linear algebra. See Theorem 6.6.6.

We can proceed even without the proof of this result: if we can find Perron numbers, then we will be able to do the computations. When the set V of nodes is small, finding Perron numbers can often be done by trial and error.

Hausdorff dimension. Now that all of the ingredients have been specified, we may proceed to analyze the Hausdorff dimension in this case. Suppose that (V, E, i, t, r) is a strongly connected, contracting Mauldin-Williams graph. Let $s > 0$ be such that s-dimensional Perron numbers q_v exist. We will suppose that the graph is strictly contracting so that $r(e) < 1$ for all e. We will compute the dimension for the path model. There is one path space $E_v^{(\omega)}$ for each node $v \in V$.

First we need metrics for the path spaces. We want the right shifts to realize similarities with the ratios assigned by the Mauldin-Williams graph. For each finite path α, let $r(\alpha)$ be the product of the numbers $r(e)$, for all the edges e in α. For $\alpha \in E_{uv}^{(*)}$, we want the diameter of $[\alpha]$ to be $r(\alpha)q_v$.

Ultrametrics ρ exist with these diameters. (One for each space $E_v^{(*)}$.) They satisfy

$$\rho(e\sigma, e\tau) = r(e)\,\rho(\sigma, \tau)$$

for $\sigma, \tau \in E_v^{(*)}$ and $e \in E_{uv}$.

Next we want to define measures on the path spaces. (The measures will all be called $\mathcal{M}$.) Because of the equations satisfied by the Perron numbers, we see that the values $(\mathrm{diam}\,[\alpha])^s$ satisfy the additivity condition (Theorem 5.5.4), namely

$$(\mathrm{diam}\,[\alpha])^s = \sum_{i(e)=t(\alpha)} (\mathrm{diam}\,[\alpha e])^s.$$

So there exists a metric measure on each of the spaces $E_v^{(\omega)}$ satisfying $\mathcal{M}([\alpha]) = (\mathrm{diam}\,[\alpha])^s$ for all $\alpha \in E_v^{(*)}$.

We can easily find an upper bound for the Hausdorff dimension. This is done in much the same way as has been done in previous cases. In terms of the maximum ratio

$$r = \max\{\,r(e) : e \in E\,\} < 1,$$

and the maximum diameter

$$q = \max\{\,q_v : v \in V\,\},$$

we can say: if α has length k, then $\mathrm{diam}\,([\alpha]) \le qr^k$. Now $qr^k \to 0$ as $k \to \infty$, so by Theorem 5.5.4, $\mathcal{H}^s(E_v^{(\omega)}) = \mathcal{M}(E_v^{(\omega)}) = q_v^s$. But $0 < q_v^s < \infty$, so $\dim E_v^{(\omega)} = s$.

Upper bound. Once we know the Hausdorff dimension of the path spaces, we can try to apply it to the sets in $\mathbf{R}^d$ that we are really interested in.

Let $(f_e)_{e \in E}$ be an iterated function system realizing the Mauldin-Williams graph (V, E, i, t, r) in $\mathbf{R}^d$. Let $(K_v)_{v \in V}$ be the unique invariant list of nonempty compact sets. As usual, we may construct the model maps

$$h_v : E_v^{(\omega)} \to \mathbf{R}^d,$$

one for each $v \in V$, such that

$$h_u(e\sigma) = f_e(h_v(\sigma)),$$

for $\sigma \in E_v^{(*)}$ and $e \in E_{uv}$. Then $h_v[E_v^{(\omega)}] = K_v$ for $v \in V$.

These maps have bounded increase, so the upper bound is easy: dim $K_v \le$ dim $E_v^{(\omega)} = s$.

Lower bound. For the lower bound, we need to limit the overlap. We will use an open set condition.

(6.4.7) Definition. If (f_e) is a realization of (V, E, i, t, r) in $\mathbf{R}^d$, then we say it satisfies the **open set condition** iff there exist nonempty open sets U_v, one for each $v \in V$, with

$$U_u \supseteq f_e[U_v]$$

for all $u, v \in V$ and $e \in E_{uv}$; and

$$f_e[U_v] \cap f_{e'}[U_{v'}] = \varnothing$$

for all $u, v, v' \in V$, $e \in E_{uv}$, $e' \in E_{uv'}$ with $e \ne e'$.

Now the argument proceeds as before. Fix a node $v \in V$. I want to show that dim $K_v \ge s$. As before, if α is a finite (nonempty) string, write α^- for its parent. Also, we will use the notation $e(x)$ for $f_e(x)$, and similarly for strings: $\alpha(x)$.

First, I claim that there is a constant $c > 0$ such that: if $A \subseteq K_v$, then the set

$$T = \Big\{ \alpha \in E_v^{(*)} : \overline{\alpha[U_{t(\alpha)}]} \cap A \ne \varnothing,$$

$$\text{diam } \alpha[U_{t(\alpha)}] < \text{diam } A \le \text{diam } \alpha^-[U_{t(\alpha-)}] \Big\}$$

has at most c elements.

If $w^* = \max_{u \in V} \text{diam } U_u$, $w_* = \min_{u \in V} \text{diam } U_u$, and $r = \min_{e \in E} r_e$, then we have for $\alpha \in T$:

$$\text{diam } \alpha[U_{t(\alpha)}] = r(\alpha) \, \text{diam } U_{t(\alpha)} \ge w_* r \, r(\alpha^-)$$

$$\ge \frac{w_* r}{w^*} \, \text{diam } \alpha^-[U_{t(\alpha-)}] \ge \frac{w_* r}{w^*} \, \text{diam } A.$$

Now if $p = \min_{u \in V} \mathcal{L}^d(U_u)$, we have the volume calculation

$$\mathcal{L}^d(\alpha[U_{t(\alpha)}]) \geq p \left(\frac{\text{diam } \alpha[U_{t(\alpha)}]}{\text{diam } U_{t(\alpha)}} \right)^d \geq p \left(\frac{w_* r}{w^{*2}} \right)^d (\text{diam } A)^d.$$

Now if $x \in A$, then every point of every set $\alpha[U_{t(\alpha)}]$ for $\alpha \in T$ is within distance $\text{diam } A + \text{diam } \alpha[U_{t(\alpha)}] < 2 \text{ diam } A$ of x. The sets $\alpha[U_{t(\alpha)}]$ are disjoint, so if T has m elements, then there are m disjoint sets, with volume at least $p(w_* r/w^{*2})^d(\text{diam } A)^d$ inside a ball with radius $2 \text{ diam } A$. If $t = \mathcal{L}^d(B_1(0))$, we have

$$mp \left(\frac{w_* r}{w^{*2}} \right)^d (\text{diam } A)^d \leq t(2 \text{ diam } A)^d.$$

Solving for m, we get

$$m \leq \frac{t}{p} \left(\frac{2w^{*2}}{w_* r} \right)^d.$$

Next, I claim that there is a constant $b > 0$ so that for any Borel set $A \subseteq K_v$, we have

$$\mathcal{M}(h_v^{-1}[A]) \leq b(\text{diam } A)^s.$$

Given A, let T be as before. Write $q = \max_{u \in V} q_u$. Then $h_v^{-1}[A] \subseteq \bigcup_{\alpha \in T}[\alpha]$. If $\alpha \in T$, then $\mathcal{M}([\alpha]) \leq r(\alpha)^s q^s \leq q^s(\text{diam } \alpha[U_{t(\alpha)}]/w_*)^s \leq q^s(\text{diam } A)^s/w_*^s$. Therefore

$$\mathcal{M}(h_v^{-1}[A]) \leq \sum_{\alpha \in T} \mathcal{M}([\alpha])$$

$$\leq \frac{cq^s}{w_*^s}(\text{diam } A)^s.$$

Then we conclude from the Method I theorem that $\mathcal{M}(h_v^{-1}[A]) \leq b\mathcal{H}^s(A)$ for all Borel sets A. In particular, $\mathcal{H}^s(K_v) \geq \mathcal{M}(h_v^{-1}[K_v])/b = \mathcal{M}(E_v^{(\omega)})/b = q_v^s/b > 0$. Therefore $\dim K_v \geq s$.

We have established the result:

(6.4.8) THEOREM. *Let (V, E, i, t, r) be a strongly connected contracting Mauldin-Williams graph describing the graph self-similarity of a list $(K_v)_{v \in V}$ of nonempty compact sets in $\mathbf{R}^d$. Let $s > 0$ be such that s-dimensional Perron numbers exist. Then $\dim K_v \leq s$ for all v. If, in addition, the realization satisfies the open set condition, then $\dim K_v = s$.*

(6.4.9) EXERCISE. *Let (V, E, i, t, r) be a strongly connected contracting Mauldin-Williams graph. Let $(f_e)_{e \in E}$ be a family of maps on $\mathbf{R}^d$ satisfying*

$$\rho(f_e(x), f_e(y)) \leq r(e)\rho(x, y).$$

Formulate the proper analog of Theorem 6.4.8.

(6.4.10)   EXERCISE. *Let* (V, E, i, t, r) *be a Mauldin-Williams graph. Let* $(f_e)_{e \in E}$ *be a family of maps on* $\mathbb{R}^d$ *satisfying* $\rho(f_e(x), f_e(y)) \geq r(e)\rho(x, y)$. *Formulate the proper analog of Theorem 6.4.8.*

Heighway's dragon boundary. We will illustrate the computation of the dimension of graph self-similar sets by the boundary K of the Heighway dragon. The graph self-similarity of K is illustrated in Figures 6.4.11, 6.4.12, and 6.4.13.

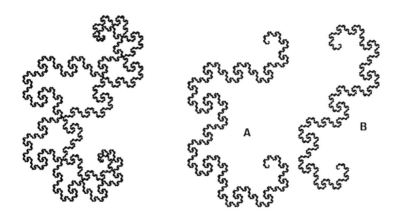

Figure 6.4.11. The Heighway dragon boundary
is made up of two parts, A and B.

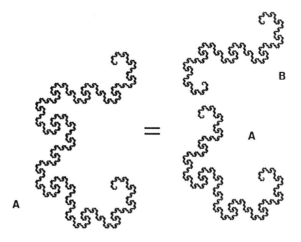

Figure 6.4.12. Set A consists of two parts.

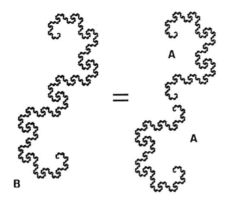

Figure 6.4.13. Set B consists of two parts.

The Mauldin-Williams graph has $V = \{\mathbf{A}, \mathbf{B}\}$ and $E = \{\mathbf{a}, \mathbf{b}, \mathbf{c}, \mathbf{d}\}$, with

$$i(\mathbf{a}) = \mathbf{A}, \quad t(\mathbf{a}) = \mathbf{A}, \quad r(\mathbf{a}) = 1/\sqrt{2},$$
$$i(\mathbf{b}) = \mathbf{A}, \quad t(\mathbf{b}) = \mathbf{B}, \quad r(\mathbf{b}) = 1/\sqrt{2},$$
$$i(\mathbf{c}) = \mathbf{B}, \quad t(\mathbf{c}) = \mathbf{A}, \quad r(\mathbf{c}) = 1/2,$$
$$i(\mathbf{d}) = \mathbf{B}, \quad t(\mathbf{d}) = \mathbf{A}, \quad r(\mathbf{d}) = 1/2.$$

(See Figure 6.4.14.)

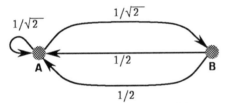

Figure 6.4.14. The Mauldin-Williams graph.

I want to find the corresponding Perron numbers. That is, I want to find positive numbers s, $q_{\mathbf{A}}$, $q_{\mathbf{B}}$ such that

$$q_{\mathbf{A}}^s = \left(\frac{1}{\sqrt{2}}\right)^s q_{\mathbf{A}}^s + \left(\frac{1}{\sqrt{2}}\right)^s q_{\mathbf{B}}^s$$
$$q_{\mathbf{B}}^s = 2\left(\frac{1}{2}\right)^s q_{\mathbf{A}}^s.$$

Let us simplify the equations by writing $\lambda = 2^{s/2}$, $x = q_{\mathbf{A}}^s$, $y = q_{\mathbf{B}}^s$; then

$$x = \frac{x}{\lambda} + \frac{y}{\lambda}$$
$$y = 2\frac{x}{\lambda^2}.$$

For a given λ, we can solve two linear equations for x and y. Of course $x = y = 0$ is always a solution. But we want $x > 0$, $y > 0$. I need to find λ so that this is possible. (In the language of linear algebra, we are solving an **eigenvalue problem**.) If there is any nonzero solution, we can multiply it by a constant and obtain another solution. So let us try to find a solution where $y = 1$. Then by the second equation we have $x = \lambda^2/2$, but also by the first equation we have

$$\lambda^3 - \lambda^2 - 2 = 0.$$

This equation for λ has one positive solution (see Figure 6.4.15), approximately 1.6956, according to Newton's method. We could write it down as a complicated expression in terms of square and cube roots, but that is less useful than the equation $\lambda^3 - \lambda^2 - 2 = 0$ and the approximate value $\lambda \approx 1.6956$.

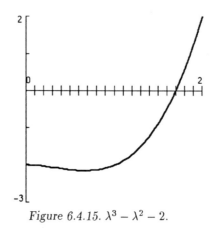

Figure 6.4.15. $\lambda^3 - \lambda^2 - 2$.

The dimension of the Mauldin-Williams graph is therefore

$$s = 2\log\lambda/\log 2 \approx 1.52,$$

and the Perron numbers may be taken as $q_A = x^{1/s} \approx 1.74$, $q_B = 1$.

In order to conclude that the Heighway dragon boundary itself has this dimension, we need to verify the open set condition.

(6.4.16) EXERCISE. *Let* $\mathbf{a}, \mathbf{b}, \mathbf{c}, \mathbf{d}$ *be the similarities realizing the two sets A and B in* $\mathbb{R}^2$. *Then there are two open sets U_A and U_B, similar to the interior of the twindragon, with*

$$\mathbf{a}[U_A] \cup \mathbf{b}[U_B] \subseteq U_A, \qquad \mathbf{a}[U_A] \cap \mathbf{b}[U_B] = \varnothing,$$
$$\mathbf{c}[U_A] \cup \mathbf{d}[U_A] \subseteq U_B, \qquad \mathbf{c}[U_A] \cap \mathbf{d}[U_A] = \varnothing.$$

Summary: The Hausdorff dimension of the boundary of Heighway's dragon is $s = 2\log\lambda/\log 2$, where λ is the real root of the equation $\lambda^3 - \lambda^2 - 2 = 0$.

(6.4.17) EXERCISE. *Compute the Hausdorff dimension of Heighway's dragon boundary using only ordinary self-similarity.*

Pentigree outline. Here is the program used to draw Figure 6.3.25.

```
; pentigree outline
make "shrink (3 + sqrt 5)/2
to pent :depth :size
    repeat 5 [A :depth :size]
end
to A :depth :size
    if :depth=0 [forward :size right 72 stop]
    B :depth-1 :size/:shrink
    A :depth-1 :size/:shrink
    A :depth-1 :size/:shrink
    BR :depth-1 :size/:shrink
end
to B :depth :size
    if :depth=0 [forward :size left 36 stop]
    C :depth-1 :size/:shrink
    A :depth-1 :size/:shrink
    BR :depth-1 :size/:shrink
end
to BR :depth :size
    if :depth=0 [forward :size left 36 stop]
    B :depth-1 :size/:shrink
    A :depth-1 :size/:shrink
    C :depth-1 :size/:shrink
end
to C :depth :size
    if :depth=0 [forward :size left 72 stop]
    B :depth-1 :size/:shrink
    BR :depth-1 :size/:shrink
end
```

(6.4.18) EXERCISE. *Does this really converge to the outline of McWorter's pentigree?*

(6.4.19) EXERCISE. *Determine the Mauldin-Williams graph describing the self-similarity of curve A.*

(6.4.20) EXERCISE. *Compute the Hausdorff dimension of the pentigree outline.*

(6.4.21) EXERCISE. *Discuss Hausdorff dimension for graph self-similar sets with Mauldin-Williams graph not strongly connected.*

Number systems. Let b be a complex number, $|b| > 1$, and let D be a finite set of complex numbers, including 0. We are interested in the numbers that can be represented in the form

$$\sum_{j=1}^{\infty} a_j b^{-j}.$$

In some cases, the set of representations may be restricted to allow only certain combinations of digits.

Consider $b = 3$ and $D = \{0, 1, 2\}$. Let F be the set of all numbers x of the form

$$x = \sum_{j=1}^{\infty} a_j b^{-j},$$

where each a_j is in the set D, and such that $a_j + a_{j+1} \leq 2$ for all j. This set is graph self-similar.

Let $F(d_1)$ be the set of numbers where the representation has $a_1 = d_1$. Let $F(d_1, d_2)$ be those numbers where the representation has $a_1 = d_1$ and $a_2 = d_2$. Then $F = F(0) \cup F(1) \cup F(2)$. The set $F(0)$ is similar to F, with ratio $1/3$. The set $F(1) = F(1, 0) \cup F(1, 1)$, since $F(1, 2) = \varnothing$. But $F(1, 0)$ is similar to F with ratio $1/9$ and $F(1, 1)$ is similar to $F(1)$ with ratio $1/3$. Finally, $F(2) = F(2, 0)$ is similar to F with ratio $1/9$. The graph is shown in Figure 6.4.22.

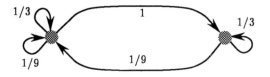

Figure 6.4.22. Graph.

(6.4.23) EXERCISE. *Compute the Hausdorff dimension of the set F.*

(6.4.24) EXERCISE. *Let $b = (1 + \sqrt{5})/2$ and $D = \{0, 1\}$. Let F be the set of numbers of the form*

$$\sum_{j=1}^{\infty} a_j b^{-j},$$

where each $a_j \in D$, and two consecutive digits 1 are not allowed. Describe the set F.

Topological dimension. The model map, which has been developed here for the purpose of computing the fractal dimension of a [graph] self-similar set, can also sometimes be used for the topological dimension as well. The model maps $h_v \colon E_v^{(\omega)} \to K$ are continuous and surjective. The spaces $E_v^{(\omega)}$ are compact. When the overlap is small, the characterization of topological dimension of Theorem 3.4.7 is often applicable. Let us do some examples.

The model map for the Cantor dust is one-to-one. Therefore the Cantor dust is zero-dimensional.

The model map for the Sierpiński gasket maps at most 2 strings to each point. Therefore the Sierpiński gasket has small inductive dimension ≤ 1. It contains line segments, so it has dimension exactly 1.

(6.4.25) Exercise. *Show that McWorter's pentigree has small inductive dimension 1.*

If you have not solved Exercise 1.6.5 yet, you can now do so painlessly.

Theorem 3.4.7 yields only an inequality. The model map for the Menger sponge (using the construction suggested in Figure 4.1.9) is 4-to-1 at some points, so we obtain the uninteresting result that the topological dimension is ≤ 3. In fact, the topological dimension is 1. Is there another way to produce the Menger sponge as a self-similar set such that the model map is only at most 2-to-one?

*6.5. Other Fractal Dimensions

According to Mandelbrot, a **fractal** is a set S with dim $S >$ ind S. We will consider (for the moment) only nonempty compact sets in Euclidean space. Mandelbrot expressed dissatisfaction with the definition for two reasons: (1) "borderline fractals" are excluded; and (2) "true geometric chaos" is included.

Borderline fractal. What might be meant by a "borderline fractal"?† This will be a set K with the usual features of fractals that we have seen often, but where dim $K =$ ind K anyway. To illustrate this, we will consider a curve, with a dragon-like construction. We begin with a sequence of positive numbers w_k, with $w_0 = 1$, $w_k > w_{k+1} > w_k/2$, and $\lim_{k\to\infty} w_k = 0$. The first set is a line segment P_0 with length $w_0 = 1$. If the polygon P_k has been constructed, consisting of many line segments of length w_k, then to construct P_{k+1}, we replace each of those line segments by two segments of length w_{k+1}. (It is possible to do this, and still have a polygon, since $w_{k+1} > w_k/2$.) If $w_k \to 0$ fast enough, we can avoid having the curve cross itself (even in the limit) by alternating between sides of the curve, as shown in the illustration. Then the limit will be homeomorphic to an interval, and therefore have small inductive dimension 1.

If we choose w_k that satisfy $w_k^s = 1/2^k$, for some $s > 1$, then this is a self-similar dragon curve that we have seen before. In the binary tree, if α is a finite string of length k, then when we use the metric and measure appropriate for the tree, we have diam $[\alpha] = w_k$ and $\mathcal{M}([\alpha]) = 1/2^k = (\text{diam}\,[\alpha])^s$. The usual calculation shows that the Hausdorff dimension for the curve will be s.

---

*An optional section.

†In fact Mandelbrot himself had in mind a much more inclusive concept: the sets of classical geometry should perhaps be included as a special case of the "fractals". We have not taken such a wide definition here.

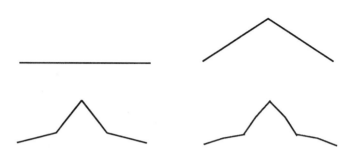

Figure 6.5.1. A dragon-like curve.

But suppose we have w_k that satisfy

$$\frac{w_k}{\log(1/w_k)} = \frac{1}{2^k}.$$

This means that w_k goes to zero more rapidly than $(1/2^k)^{1/s}$ for any $s > 1$, but more slowly than $1/2^k$ itself.

(6.5.2) EXERCISE. *When*

$$\frac{w_k}{\log(1/w_k)} = \frac{1}{2^k},$$

the topological and Hausdorff dimensions are both 1.

Why would we call this a "borderline fractal"?

Packing dimension. Mandelbrot says that his definition for "fractal" (ind $S <$ dim S) is too broad, in that it admits "true geometric chaos". The sets that are of interest for applications (and in mathematics) are generally not the most general set, with few special properties. So it may be useful to restrict the term "fractal" so that the sets meeting the conditions have useful properties. One possible way to do this has been proposed by James Taylor.

Taylor introduced another fractal dimension, called the "packing dimension". He showed that several of the fractal dimensions always have values that are between the packing dimension and the Hausdorff dimension (see Exercise 6.5.15, below). He proposed to apply the term "fractal" to (nonempty compact) sets where the packing dimension is equal to the Hausdorff dimension (and therefore to all the dimensions in between), but not equal to the topological dimension.

The idea for the packing measure of a set F is to try to arrange as many disjoint balls as possible, with centers in F. If $\varepsilon > 0$, and $s > 0$, let

$$\widetilde{\mathcal{P}}_\varepsilon^s(F) = \sup \sum_{B \in \mathcal{D}} (\operatorname{diam} B)^s,$$

where the sup is over all countable collections $\mathcal{D}$ of disjoint balls with center in F and diameter $\leq \varepsilon$. Then, when ε decreases to 0, this value decreases, so we define

$$\widetilde{\mathcal{P}}^s(F) = \lim_{\varepsilon \to 0} \widetilde{\mathcal{P}}_\varepsilon^s(F) = \inf_{\varepsilon > 0} \widetilde{\mathcal{P}}_\varepsilon^s(F).$$

When we have done this, we get a family $\left(\widetilde{\mathcal{P}}^s\right)$ of set functions indexed by s. As before, there is a critical value:

(6.5.3) EXERCISE. *Let F be a set in a metric space. There is $s_0 \in [0, \infty]$ such that*

$$\widetilde{\mathcal{P}}^s(F) = \infty \qquad \text{for all } s < s_0;$$
$$\widetilde{\mathcal{P}}^s(F) = 0 \qquad \text{for all } s > s_0.$$

This critical value s_0 will be called the **packing index** of the set F. However, the set functions $\widetilde{\mathcal{P}}^s$ are not really what we want. They are *not* outer measures. This is not unexpected, since the process used to construct them is not method II. Here is an illustration showing that $\widetilde{\mathcal{P}}^{1/2}$ fails to be an outer measure on $\mathbf{R}$:

(6.5.4) PROPOSITION. *Let K be the compact set $\{0, 1, 1/2, 1/3, 1/4, 1/5, \cdots\} \subseteq \mathbf{R}$. Then $\widetilde{\mathcal{P}}^{1/2}(K) > 0$.*

PROOF: Let $k \in \mathbf{N}$ be odd, let $\varepsilon = 2^{-k}$, and let $n = 2^{(k-1)/2}$. Then

$$\frac{1}{n-1} - \frac{1}{n} > \frac{1}{n^2} = 2\varepsilon,$$

so the balls with radius ε and centers $1, 1/2, 1/3, \cdots, 1/n$ are disjoint. So

$$\widetilde{\mathcal{P}}_{2\varepsilon}^{1/2}(K) \geq n \, (2\varepsilon)^{1/2} = 1,$$

and therefore $\widetilde{\mathcal{P}}^{1/2}(K) \geq 1$. ☺

It seems unreasonable to claim that this countable set K has positive dimension. We know a good way (method I) to get an outer measure from a set function. So we apply method I to the set function $\widetilde{\mathcal{P}}^s$:

$$\overline{\mathcal{P}}^s(F) = \inf \sum_{C \in \mathcal{C}} \widetilde{\mathcal{P}}^s(C),$$

where the inf is over all countable covers $\mathcal{C}$ of the set F. Note that $\widetilde{\mathcal{P}}^s(C) = \widetilde{\mathcal{P}}^s\left(\overline{C}\right)$, so we may restrict consideration to covers $\mathcal{C}$ of F consisting of closed sets.

(6.5.5) EXERCISE. *The set function $\overline{\mathcal{P}}^s$ is a metric outer measure.*

The restriction of $\overline{\mathcal{P}}^s$ to the measurable sets is a measure, called the **s-dimensional packing measure**, and written $\mathcal{P}^s$. As usual there is a critical value for s:

(6.5.6) EXERCISE. *Let F be a Borel set in a metric space. There is $s_0 \in [0, \infty]$ such that*

$$\mathcal{P}^s(F) = \infty \qquad \text{for all } s < s_0;$$
$$\mathcal{P}^s(F) = 0 \qquad \text{for all } s > s_0.$$

This value s_0 is called the **packing dimension** of the set F. We will write $s_0 = \operatorname{Dim} F$. It is a more reasonable quantity than the packing index defined above.

The packing dimension is related to the Hausdorff dimension.

(6.5.7) PROPOSITION. *For any Borel set $F \subseteq \mathbb{R}^d$, we have $\dim F \le \operatorname{Dim} F$.*

PROOF: I first show that $\overline{\mathcal{H}}^s_{2\varepsilon}(F) \le 2^s \widetilde{\mathcal{P}}^s_\varepsilon(F)$. Now if $\widetilde{\mathcal{P}}^s_\varepsilon(F) = \infty$, then this is clear. So suppose $\widetilde{\mathcal{P}}^s_\varepsilon(F) < \infty$. If there is an *infinite* family of disjoint balls with diameter ε and centers in F, then $\widetilde{\mathcal{P}}^s_\varepsilon(F) = \infty$. So there is a maximal finite family $\{\, B_{\varepsilon/2}(x_i) : 1 \le i \le n \,\}$ of disjoint balls with $x_i \in F$; that is, for any $x \in F$, there is some i between 1 and n with $B_{\varepsilon/2}(x) \cap B_{\varepsilon/2}(x_i) \ne \varnothing$. We have

$$\widetilde{\mathcal{P}}^s_\varepsilon(F) \ge \sum_{i=1}^{n} \big(\operatorname{diam} B_{\varepsilon/2}(x_i)\big)^s = n\varepsilon^s.$$

But the collection with double diameters, $\{\, B_\varepsilon(x_i) : 1 \le i \le n \,\}$, covers F, so

$$\overline{\mathcal{H}}^s_{2\varepsilon}(F) \le \sum_{i=1}^{n} \big(\operatorname{diam} B_\varepsilon(x_i)\big)^s = n(2\varepsilon)^s.$$

Therefore $\overline{\mathcal{H}}^s_{2\varepsilon}(F) \le 2^s \widetilde{\mathcal{P}}^s_\varepsilon(F)$.

Now take the limit as $\varepsilon \to 0$ and conclude $\mathcal{H}^s(F) \le 2^s \widetilde{\mathcal{P}}^s(F)$. So by the Method I theorem, $\mathcal{H}^s(F) \le 2^s \mathcal{P}^s(F)$.

Now if $s < \dim F$, then $\mathcal{H}^s(F) = \infty$, so $\mathcal{P}^s(F) = \infty$, and therefore $s \le \operatorname{Dim} F$. We therefore conclude that $\dim F \le \operatorname{Dim} F$. ☺

A set $F \subseteq \mathbb{R}^d$ is a **fractal** (in the sense of Taylor) iff $\operatorname{ind} F < \dim F = \operatorname{Dim} F$.

(6.5.8) EXERCISE. *If F is a Borel subset of a metric space (not $\mathbb{R}^d$), then does it follow that $\dim F \le \operatorname{Dim} F$?*

Packing dimension and self-similarity. The self-similar sets K that we have considered before satisfy the condition $\dim K = \operatorname{Dim} K$. The machinery for the proof of this fact has already been developed.

Let $(r_e)_{e \in E}$ be a contracting ratio list. Let s be its dimension. The string model is $E^{(\omega)}$ as usual. Recall that a metric ρ and a measure $\mathcal{M}$ exist on $E^{(\omega)}$ satisfying $\mathcal{M}([\alpha]) = \big(\operatorname{diam}[\alpha]\big)^s$.

(6.5.9) THEOREM. *Let* s *be the similarity dimension of* $E^{(\omega)}$. *Then* $\mathcal{M}(E) = \mathcal{P}^s(E)$ *for all Borel sets* E. *Therefore the packing dimension of* $E^{(\omega)}$ *is equal to* s.

PROOF: The balls in $E^{(\omega)}$ are exactly the sets $[\alpha]$, where $\alpha \in E^{(*)}$. Every point of a ball is a center of the ball (Exercise 2.1.15). Fix a finite string $\beta \in E^{(*)}$. Let $\varepsilon > 0$ be such that $\varepsilon < \operatorname{diam}[\beta]$. Choose k so that $\operatorname{diam}[\alpha] < \varepsilon$ for all $\alpha \in E^{(k)}$. Then $\mathcal{D} = \{ [\alpha] : \alpha \in E^{(k)}, \alpha \geq \beta \}$ is a collection of disjoint balls with centers in $[\beta]$ and diameter $\leq \varepsilon$. So $\widetilde{\mathcal{P}}_\varepsilon^s([\beta]) \geq \sum_{\mathcal{D}} (\operatorname{diam}[\alpha])^s = \sum_{\mathcal{D}} \mathcal{M}([\alpha]) = \mathcal{M}([\beta])$. On the other hand, let $\mathcal{D}$ be any collection of disjoint balls with centers in $[\beta]$, and diameters $\leq \varepsilon$. Now the distance from any point of $[\beta]$ to any point of the complement is $> \varepsilon$, so $B \subseteq [\beta]$ for all $B \in \mathcal{D}$. Then $\sum_{B \in \mathcal{D}} (\operatorname{diam} B)^s = \sum_{B \in \mathcal{D}} \mathcal{M}(B) \leq \mathcal{M}([\beta])$. Therefore $\widetilde{\mathcal{P}}^s([\beta]) = \mathcal{M}([\beta])$.

Now suppose U is any open set. Then it may be written as a disjoint union $U = \bigcup_{i \in \mathbb{N}} [\alpha_i]$. So $\overline{\mathcal{P}}^s(U) \leq \sum \widetilde{\mathcal{P}}^s([\alpha_i]) = \sum \mathcal{M}([\alpha_i]) = \mathcal{M}(U)$.

Now let K be a compact set. Given $\varepsilon > 0$, choose k so that $\operatorname{diam}[\alpha] < \varepsilon$ for all $\alpha \in E^{(k)}$. Then $\mathcal{D} = \{ [\alpha] : |\alpha| = k, [\alpha] \cap K \neq \varnothing \}$ is a finite disjoint family of balls, with diameters $< \varepsilon$, that covers K. So $\widetilde{\mathcal{P}}_\varepsilon^s(K) \geq \sum_{B \in \mathcal{D}} (\operatorname{diam} B)^s = \sum_{B \in \mathcal{D}} \mathcal{M}(B) = \mathcal{M}(\bigcup_{B \in \mathcal{D}} B) \geq \mathcal{M}(K)$. So $\widetilde{\mathcal{P}}^s(K) \geq \mathcal{M}(K)$.

Now let E be any $\mathcal{M}$-measurable set. Consider closed sets K_i covering E. Then by the above, $\sum \widetilde{\mathcal{P}}^s(K_i) \geq \sum \mathcal{M}(K_i) \geq \mathcal{M}(\bigcup K_i) \geq \mathcal{M}(E)$. Therefore $\overline{\mathcal{P}}^s(E) \geq \mathcal{M}(E)$. On the other hand, given $\varepsilon > 0$, by 5.2.7 there is an open set $U \supseteq E$ with $\mathcal{M}(U \setminus E) < \varepsilon$. So $\overline{\mathcal{P}}^s(E) \leq \overline{\mathcal{P}}^s(U) \leq \mathcal{M}(U) \leq \mathcal{M}(E) + \varepsilon$. Therefore $\overline{\mathcal{P}}^s(E) = \mathcal{M}(E)$. ☺

(6.5.10) THEOREM. *Let* $(f_e)_{e \in E}$ *be a realization of the contracting ratio list* (r_e) *in some Euclidean space* $\mathbb{R}^d$. *The packing dimension* $\operatorname{Dim} K$ *of the invariant set* K *is not greater than the similarity dimension* s. *If the open set condition is satisfied, then* $\operatorname{Dim} K = s$.

PROOF: The model map $h \colon E^{(\omega)} \to \mathbb{R}^d$ has bounded increase. So there is a constant $b > 0$ with $|h(\sigma) - h(\tau)| \leq b \rho(\sigma, \tau)$.

In order to use this fact, we will need an estimate of the diameter of a ball $B_\varepsilon(\sigma)$. If $\varepsilon < 1$, then $B_\varepsilon(\sigma) = [\sigma{\upharpoonright}k]$, where k is chosen so that we have $\operatorname{diam}[\sigma{\upharpoonright}k] < \varepsilon \leq \operatorname{diam}[\sigma{\upharpoonright}(k-1)]$. Thus, if $r = \min_e r_e$, then we have $\operatorname{diam} B_\varepsilon(\sigma) = \operatorname{diam}[\sigma{\upharpoonright}k] \geq r \operatorname{diam}[\sigma{\upharpoonright}(k-1)] \geq r\varepsilon$.

Now let $\varepsilon > 0$ and $\varepsilon < b$. Let $\{B_{\varepsilon_i}(x_i)\}_{i=1}^\infty$ be a countable disjoint family of balls with $x_i \in K$ and $\varepsilon_i < \varepsilon/2$. Choose $\sigma_i \in E^{(\omega)}$ with $h(\sigma_i) = x_i$. Then the balls $B_{\varepsilon_i/b}(\sigma_i)$ are disjoint, since $h[B_{\varepsilon_i/b}(\sigma_i)] \subseteq B_{\varepsilon_i}(x_i)$. They have diameters $\leq \varepsilon_i/b < 1$. Therefore $\sum_i (\operatorname{diam} B_{\varepsilon_i}(x_i))^s \leq \sum_i (2\varepsilon_i)^s = (2b/r)^s \sum_i (r\varepsilon_i/b)^s \leq (2b/r)^s \sum_i (\operatorname{diam} B_{\varepsilon_i/b}(\sigma_i))^s \leq (2b/r)^s \mathcal{M}(E^{(\omega)}) = (2b/r)^s$. So $\widetilde{\mathcal{P}}_\varepsilon^s(K) \leq (2b/r)^s$ for all ε, so $\widetilde{\mathcal{P}}^s(K) \leq (2b/r)^s$, and therefore $\mathcal{P}^s(K) \leq (2b/r)^s$. This is finite, so $\operatorname{Dim} K \leq s$.

Now suppose the open set condition is satisfied. By Theorem 6.3.12, we have $s \le \dim K \le \operatorname{Dim} K$, so $\operatorname{Dim} K = s$. ☺*

(6.5.11) EXERCISE. *Suppose $K \subseteq \mathbf{R}^d$ has graph self-similarity for a contracting, strongly connected Mauldin-Williams graph. If s-dimensional Perron numbers exist, then $\operatorname{Dim} K \le s$. If, in addition, the open set condition holds, then $\operatorname{Dim} K = s$.*

Entropy indices. We will next discuss some other fractal dimensions known as entropy indices. We will consider only $\mathbf{R}^2$ for simplicity.

If $r > 0$, then the **square net** of side r consists of all squares of the form

$$A = \big[(m-1)r, mr\big) \times \big[(n-1)r, nr\big),$$

where $m, n \in \mathbf{Z}$. Write $\mathcal{S}_r$ for this set of squares. So the plane $\mathbf{R}^2$ is the disjoint union of this countable collection of squares. Write $\mathcal{S} = \bigcup_{r > 0} \mathcal{S}_r$.

For $s > 0$, consider the method II outer measure $\overline{\mathcal{M}}^s$ defined using the set function $C \colon \mathcal{S} \to [0, \infty)$ defined by: $C(A) = r^s$ if $A \in \mathcal{S}_r$. Now any set $B \subseteq \mathbf{R}^2$ of diameter r is contained in the union of at most 4 sets of $\mathcal{S}_r$. On the other hand, a square with side r has diameter $\sqrt{2}\,r$. This means that

$$2^{-s/2}\,\overline{\mathcal{H}}^s(F) \le \overline{\mathcal{M}}^s(F) \le 4\,\overline{\mathcal{H}}^s(F).$$

Therefore $s_0 = \dim F$ is the critical value for which

$$\overline{\mathcal{M}}^s(F) = \infty \qquad \text{for all } s < s_0;$$
$$\overline{\mathcal{M}}^s(F) = 0 \qquad \text{for all } s > s_0.$$

Some calculations involving the Hausdorff dimension are simplified by using this alternative to the definition (for example [20, Chapter 5]).

Now we discuss another variant. Fix a number $r > 0$, and cover only by sets of $\mathcal{S}_r$; then let $r \to 0$. It should be emphasized that this is not method II. Now if $\mathcal{A} \subseteq \mathcal{S}_r$ covers a set F, then $\sum_{A \in \mathcal{A}} r^s$ is just $N r^s$, where N is the number of elements of $\mathcal{A}$. So the definition may be phrased as follows. Let $N_r(F)$ be the number of sets of the square net $\mathcal{S}_r$ that intersect F. Define

$$\widetilde{\mathcal{K}}_r^s(F) = N_r(F) \cdot r^s,$$
$$\widetilde{\mathcal{K}}^s(F) = \liminf_{r \to 0} \widetilde{\mathcal{K}}_r^s(F).$$

As usual there is a critical value for s.

---

*In fact, we have proved that the *packing index* is also s.

(6.5.12) EXERCISE. *Let*

$$s_0 = \liminf_{r \to 0} \frac{\log N_r(F)}{\log(1/r)}.$$

Then

$$\widetilde{\mathcal{K}}^s(F) = \infty \qquad \text{for all } s < s_0;$$
$$\widetilde{\mathcal{K}}^s(F) = 0 \qquad \text{for all } s > s_0.$$

The critical value s_0 will be called the **lower entropy index** of F. It is also known as the **box dimension**, or **box-counting dimension**. The set functions $\widetilde{\mathcal{K}}^s$ have the same shortcoming as $\widetilde{\mathcal{P}}^s$.

(6.5.13) EXERCISE. *There is a countable compact set K with positive lower entropy index.*

A variant is obtained by replacing lim inf with lim sup. The set functions are then

$$\limsup_{r \to 0} \widetilde{\mathcal{K}}_r^s(F).$$

The critical value for s, called the the **upper entropy index**,† is given by

$$\limsup_{r \to 0} \frac{\log N_r(F)}{\log(1/r)}.$$

(6.5.14) EXERCISE. *Let $F \subseteq \mathbb{R}^2$. The Hausdorff dimension of F is not larger than the lower entropy index of F, which is not larger than the upper entropy index of F, which is not larger than the packing index of F.*

Since the set functions involved are not outer measures, it is customary to apply method I to them. Let

$$\overline{\mathcal{K}}^s(F) = \inf \sum_{C \in \mathcal{D}} \widetilde{\mathcal{K}}^s(C),$$

where the inf is over all countable covers $\mathcal{D}$ of the set F. This is a metric outer measure. The critical value is the **lower entropy dimension** of F. A similar definition can be given for the **upper entropy dimension**.

(6.5.15) EXERCISE. *The Hausdorff dimension dim F of F is not larger than the lower entropy dimension of F, which is not larger than the upper entropy dimension of F, which is not larger than the packing dimension Dim F of F.*

So if F is a fractal in the sense of Taylor, these four fractal dimensions all coincide.

(6.5.16) EXERCISE. *Define the set functions $\widetilde{\mathcal{K}}^s$ and $\overline{\mathcal{K}}^s$ in the space $\mathbb{R}^d$. Prove analogs of 6.5.14 and 6.5.15.*

(6.5.17) EXERCISE. *Define lower entropy index and lower entropy dimension on a general metric space S in such a way that the definitions agree in $\mathbb{R}^2$ with the definitions already given.*

---

†Barnsley [3] uses the term "fractal dimension" for this value.

*6.6. REMARKS

Felix Hausdorff [27] formulated the concepts used in this chapter: today we call them the Hausdorff measures and the Hausdorff dimension. Almost all of the early work on the subject was done by A. N. Besicovitch [4]. Mandelbrot therefore uses the term "Hausdorff-Besicovitch dimension".

In fact, Hausdorff proposed a much more general class of measures than the ones discussed here. For example, he proposed using functions of the diameter other than a power: for example, the function $h(x) = x^s \left(1/\log(1/x)\right)^t$ corresponds to the construction in 6.5.2 when $s = t = 1$. He also proposed using characteristics of the covering sets other than the diameter. The seminal paper [27] is required reading for the aspiring expert on fractals.

Computation of the Hausdorff dimension using self-similarity appears even in Hausdorff's paper. It was carefully worked out by P. A. P. Moran [42] for subsets of $\mathbb{R}$, and by John Hutchinson [31] for subsets of $\mathbb{R}^d$. The open set condition is used in both of these papers.

The generalization of self-similarity that we have called "graph self-similarity" has a complicated history. The version that is used here is based on the work of R. Daniel Mauldin and S. C. Williams [39]. The "two-part dust" was invented explicitly to illustrate the computation of the Hausdorff dimension for graph self-similar sets.

For the packing dimension, see [49] and [50], for example. Fractal dimensions in addition to those defined here can be found in [36, p. 357*ff*] and [37].

The pentadendrite was shown to me by my colleague W. A. McWorter. The terdragon comes from Chandler Davis and Donald Knuth [11].

Topological vs. Hausdorff dimension. The small inductive dimension of a space S is a topological property of the space. That is, if S is homeomorphic to T, then ind S = ind T. The Hausdorff dimension is not a topological property. The spaces $(E^{(\omega)}, \rho_r)$, where $E = \{0, 1\}$ is a two-letter alphabet, are all homeomorphic, but the Hausdorff dimension varies as r varies. We know that ind $S \leq$ dim S. In fact, ind S is the largest topologically invariant lower bound for dim S:

(6.6.1) THEOREM. *Let S be a separable metric space. Then*

$$\text{ind } S = \inf \left\{ \dim T : T \text{ is homeomorphic to } S \right\}.$$

I will prove here only the simplest case:

---

*An optional section.

(6.6.2) PROPOSITION. *Let S be a separable zero-dimensional metric space. Then*

$$0 = \inf \{ \dim T : T \text{ is homeomorphic to } S \} .$$

PROOF: First, S is homeomorphic to a subset T of the string space $\{\mathbf{0},\mathbf{1}\}^{(\omega)}$, by Theorem 3.1.9. With metric ρ_r, the space $\{\mathbf{0},\mathbf{1}\}^{(\omega)}$ has Hausdorff dimension $\log 2/\log(1/r)$. But $\lim_{r \to 0} \log 2/\log(1/r) = 0$. ☺

The general case may be proved in a similar way [**30**, Theorem VII 5]. For example, the Menger sponge is a universal 1-dimensional space, so metric spaces homeomorphic to the Menger sponge, but with Hausdorff dimension close to 1 should be exhibited. The approximation shown in Figure 6.6.3 suggests the idea. It is self-affine, rather than self-similar, so our methods of computation will not evaluate its Hausdorff dimension, however.

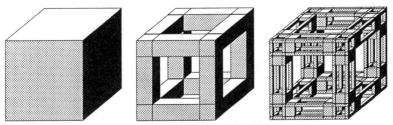

Figure 6.6.3. Homeomorph of the Menger sponge.

Two-dimensional Lebesgue measure compared to two-dimensional Hausdorff measure. Now we will discuss the assertion: In $\mathbf{R}^2$, there is a constant factor b so that $\mathcal{H}^2 = b\mathcal{L}^2$. In Theorem 6.5.2, we saw that

$$a\mathcal{L}^2(B) \leq \mathcal{H}^2(B) \leq b\mathcal{L}^2(B),$$

where $a = 1/\pi$ and $b = \mathcal{H}^2(Q)$. (Q is the unit square.)

To improve the lower bound, we need an interesting fact from two-dimensional geometry: Among all sets with a given diameter, the disk has the largest area. That is, if A is a set with diameter t, then $\overline{\mathcal{L}}^2(A) \leq \pi t^2/4$.† The proof requires some knowledge concerning convexity in two dimensions. First, A has the same diameter as its convex hull, so we may assume that A is convex. Similarly we may assume that A is closed. Choose any boundary point of A; let it be the origin of coordinates. A has a support line there, let it be the x-axis. Then the set A is given in polar coordinates (r, θ) by equations

$$0 \leq r \leq R(\theta), \quad 0 \leq \theta \leq \pi,$$

---

†The corresponding fact for higher dimensions can be proved from Steiner's symmetrization construction. See, for example, [**16**, p. 107].

for some function R. Now A has diameter t. So the distance between the polar points $(R(\theta), \theta)$ and $(R(\theta + \pi/2), \theta + \pi/2)$ is at most t (Figure 6.6.4). By the Pythagorean theorem, we may conclude

$$R(\theta)^2 + R(\theta + \pi/2)^2 \le t^2.$$

Then the area may be computed in polar coordinates:

$$\int_0^\pi \frac{R(\theta)^2}{2} \, d\theta = \int_0^{\pi/2} \frac{R(\theta)^2}{2} \, d\theta + \int_{\pi/2}^\pi \frac{R(\theta)^2}{2} \, d\theta$$

$$= \int_0^{\pi/2} \frac{R(\theta)^2 + R(\theta + \pi/2)^2}{2} \, d\theta$$

$$\le \int_0^{\pi/2} \frac{t^2}{2} \, d\theta = \frac{\pi t^2}{4}.$$

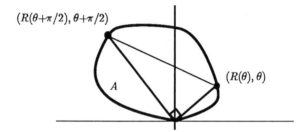

Figure 6.6.4. Polar coordinates.

So: a set $A \subseteq \mathbb{R}^2$ with diameter t has area at most $\pi t^2/4$. Then the argument given in Theorem 6.2.5, with $a = 4/\pi$, will show that $(4/\pi)\mathcal{L}^2(B) \le \mathcal{H}^2(B)$ for any Borel set B.

For the upper bound, we use the Vitali covering theorem [**19**, Theorem 1.10]. Let $b = \mathcal{H}^2(Q)$, where Q is the open unit square. Now a disk can be approximated inside and outside by little squares, so we have (by the argument in the proof of Theorem 6.2.5) $\mathcal{H}^2(B) = b\mathcal{L}^2(B)$ for all disks B. The collection of all closed disks with diameter $< \varepsilon$ and contained in the square Q satisfies the hypothesis of the Vitali theorem, so there is a countable disjoint set $\{ B_i : i \in \mathbb{N} \}$ of them with $\mathcal{L}^2(Q \setminus \bigcup_{i \in \mathbb{N}} B_i) = 0$. But then, by the inequality $\mathcal{H}^2(B) \le b\mathcal{L}^2(B)$, we know that $\mathcal{H}^2(Q \setminus \bigcup_{i \in \mathbb{N}} B_i)$ is also 0, so $\mathcal{H}^2_\varepsilon(Q \setminus \bigcup_{i \in \mathbb{N}} B_i) = 0$. Now

$$\mathcal{H}^2_\varepsilon \left(\bigcup_{i \in \mathbb{N}} B_i \right) \leq \sum_{i=1}^{\infty} (\text{diam } B_i)^2$$

$$= (4/\pi) \sum_{i=1}^{\infty} (\pi/4)(\text{diam } B_i)^2$$

$$= (4/\pi) \sum_{i=1}^{\infty} \mathcal{L}^2(B_i)$$

$$= (4/\pi)\mathcal{L}^2(Q) = 4/\pi.$$

Therefore $b = \mathcal{H}^2(Q) = \mathcal{H}^2(\bigcup_{i \in \mathbb{N}} B_i) \leq 4/\pi$.

So we have exactly $\mathcal{H}^2(B) = (4/\pi)\mathcal{L}^2(B)$ for all Borel sets B.

The same result is true in $\mathbb{R}^d$, namely $\mathcal{H}^d(B) = b_d \mathcal{L}^d(B)$, where b_d is the appropriate constant $1/\mathcal{L}^d(B_{1/2}(0))$.

The dimension of a Mauldin-Williams graph. The dimension of a Mauldin-Williams graph exists and is unique. The proof of this fact will be given here. It requires some knowledge of linear algebra. In particular, it requires information from the Perron-Frobenius theorem (stated below).

Let A be a square matrix. The **spectral radius** of A is the maximum of the absolute values of all of the complex eigenvalues of A. We will write $r(A)$ for the spectral radius of A.

We will use some additional notation: $A \geq 0$ means all of the entries of A are nonnegative, and $A > 0$ means all of the entries of A are positive; $A \geq B$ means $A - B \geq 0$, and $A > B$ means $A - B > 0$. The matrix $A \geq 0$ is called **reducible** iff the rows and columns can be permuted (by the same permutation) so that A has the form

$$A = \begin{bmatrix} B & O \\ C & D \end{bmatrix},$$

where B and D are square matrices (with at least one row each), and O is a rectangular matrix of zeros. If A is not reducible, then it is **irreducible**. A column matrix with all entries 0 is **0**, and a column matrix with all entries 1 is **1**.

Here is (part of) the Perron-Frobenius theorem. See [**22**, Chapter XIII] for a proof.

(6.6.5) THEOREM. *Let $A \geq 0$ be an irreducible square matrix, and let $\lambda \in \mathbb{R}$. Then:*

(1) *If $\lambda = r(A)$, then there is a column matrix $\mathbf{x} > 0$ with $A\mathbf{x} = \lambda\mathbf{x}$.*
(2) *If there is a nonzero column matrix $\mathbf{x} \geq 0$ with $A\mathbf{x} = \lambda\mathbf{x}$, then $\lambda = r(A)$.*
(3) *If there is a nonzero column matrix $\mathbf{x} \geq 0$ with $A\mathbf{x} < \lambda\mathbf{x}$, then $\lambda > r(A)$.*
(4) *If there is a nonzero column matrix $\mathbf{x} \geq 0$ with $A\mathbf{x} > \lambda\mathbf{x}$, then $\lambda < r(A)$.*

Now we are in a position to prove that the dimension of a strongly connected, strictly contracting Mauldin-Williams graph exists and is unique.

(6.6.6) THEOREM. *Let (V, E, i, t, r) be a strongly connected, strictly contracting Mauldin-Williams graph. There is a unique number $s \geq 0$ such that positive numbers q_v exist satisfying*

$$q_u^s = \sum_{\substack{v \in V \\ e \in E_{uv}}} r(e)^s \cdot q_v^s$$

for all $u \in V$.

PROOF: We will be using matrices with the rows (and columns) labeled by V. For each pair $u, v \in V$, and $s \geq 0$, let

$$A_{uv}(s) = \sum_{e \in E_{uv}} r(e)^s.$$

Let $A(s)$ be the matrix with entry $A_{uv}(s)$ in row u column v. Let $\Phi(s) = r(A(s))$ be the spectral radius of the matrix $A(s)$. Now the matrix $A(s)$ has nonnegative entries. The entry $A_{uv}(s)$ is positive if and only if E_{uv} is not empty. Since the graph is strongly connected, the matrices $A(s)$ are irreducible. I will prove: (a) s-dimensional Perron numbers exist if and only if $\Phi(s) = 1$, and (b) the equation $\Phi(s) = 1$ has a unique solution in $[0, \infty)$.

First, suppose that s-dimensional Perron numbers exist, so that

$$q_u^s = \sum_{\substack{v \in V \\ e \in E_{uv}}} r(e)^s \cdot q_v^s$$

for all $u \in V$. Thus, if the column matrix $\mathbf{x}$ has entries q_v^s, then $\mathbf{x} > 0$ and $A(s)\mathbf{x} = \mathbf{x}$, so by the Perron-Frobenius theorem, 1 is the spectral radius of $A(s)$.

Conversely, suppose that $1 = r(A(s))$. Then by the Perron-Frobenius theorem, there is a column matrix $\mathbf{x} > 0$ with $A(s)\mathbf{x} = \mathbf{x}$. If we write x_v for the entries of $\mathbf{x}$, then the numbers $q_v = x_v^{1/s}$ will be s-dimensional Perron numbers.

Next, I claim that the function Φ is continuous. Certainly the entries $A_{uv}(s)$ of the matrix are continuous functions of s. Fix a number s_0. let $\mathbf{x} > 0$ be the Perron-Frobenius eigenvector: $A(s_0)\mathbf{x} = \Phi(s_0)\mathbf{x}$. Let the entries of $\mathbf{x}$ be x_v. Define positive numbers a, b by $a = \min x_v$, $b = \max x_v$. Suppose V has n elements, so the matrices are $n \times n$. Let $\varepsilon > 0$ be given. By the continuity of the entries A_{uv}, there exists $\delta > 0$ so that if $|s - s_0| < \delta$, then

$$|A_{uv}(s) - A_{uv}(s_0)| < \frac{a\varepsilon}{nb}$$

for all u, v. Now we have

$$\sum_v A_{uv}(s)x_v = \sum_v A_{uv}(s_0)x_v + \sum_v \big(A_{uv}(s) - A_{uv}(s_0)\big)x_v$$
$$\leq \Phi(s_0)x_u + n\frac{a\varepsilon}{nb}b$$
$$\leq \big(\Phi(s_0) + \varepsilon\big)x_u.$$

Therefore, by the Perron-Frobenius theorem, $\Phi(s) = r(A(s)) \leq \Phi(s_0) + \varepsilon$. Similarly $\Phi(s) \geq \Phi(s_0) - \varepsilon$. This shows that Φ is continuous.

Since the graph is strongly connected, each row has at least one nonzero entry. So for each u there is v with $A_{uv}(0) \geq 1$. Therefore $A(0)\mathbf{1} \geq \mathbf{1}$, so that $\Phi(0) \geq 1$. The entries $A_{uv}(s) \to 0$ as $s \to \infty$, so for large enough s, we have $A_{uv}(s) \leq 1/(2n)$ for all u, v, so that $A(s)\mathbf{1} \leq (1/2)\mathbf{1}$, and thus $\Phi(s) \leq 1/2$. Now by the intermediate value theorem, there is a solution s to the equation $\Phi(s) = 1$.

Finally, to prove the uniqueness, we will show that Φ is strictly decreasing. The derivative of A_{uv} is ≤ 0, and in fact < 0 unless A_{uv} is identically 0. Each row has at least one nonzero entry, so if $s > s_0$ and $\mathbf{x}$ is the Perron-Frobenius eigenvector for $A(s_0)$, we have $A(s)\mathbf{x} < A(s_0)\mathbf{x} = \Phi(s_0)\mathbf{x}$. So $\Phi(s) < \Phi(s_0)$. Therefore the function Φ is strictly decreasing. ☺

(6.6.7) EXERCISE. *Let (V, E, i, t, r) be a contracting, strongly connected Mauldin-Williams graph. Are the conclusions of Theorem 6.6.6 still correct?*

To compute the dimension of a strictly contracting, strongly connected Mauldin-Williams graph, we would ordinarily find the numbers s such that 1 is an eigenvalue of the matrix $A(s)$. If that s is unique, it is the dimension. If not, then we find for each s the corresponding eigenvector for $A(s)$; only one of the values s will admit an eigenvector with all entries positive.

Remarks on the exercises. Exercise 6.1.9. Suppose $f : S \to T$ has bounded increase, and $A \subseteq S$. Say $\rho(f(x), f(y)) \leq b\rho(x, y)$. If $\mathcal{D}$ is a countable cover of A by sets with diameter at most ε, then $\mathcal{D}' = \{ f[D] : D \in \mathcal{D} \}$ is a countable cover of $f[A]$ by sets with diameter at most $b\varepsilon$. Now

$$\sum_{D \in \mathcal{D}} (\operatorname{diam} f[D])^s \leq b^s \sum_{D \in \mathcal{D}} (\operatorname{diam} D)^s,$$

so $\overline{\mathcal{H}}^s_{b\varepsilon}(f[A]) \leq b^s \overline{\mathcal{H}}^s_{\varepsilon}(A)$. Therefore $\dim f[A] \leq \dim A$. The case of bounded decrease is similar.

Exercise 6.2.11: Suppose $\operatorname{ind} S \geq 1$. Then S does not have a base for the open sets consisting of clopen sets. So there is a point $a \in S$ and $\varepsilon > 0$ such that for $0 < r < \varepsilon$, the ball $B_r(a)$ is not clopen. The function $h : S \to \mathbb{R}$ defined by $h(x) = \rho(x, a)$ satisfies

$$|h(x) - h(y)| \leq \rho(x, y).$$

Its range includes the interval $(0, \varepsilon)$. So we have $\dim S \geq \dim h[S] \geq \dim (0, \varepsilon) = 1$.

Exercise 6.3.15: Marion's open set condition is from [38].

For Exercise 6.3.17: The terdragon boundary is made up of two copies of the 120-degree dragon of Figure 1.5.8. The open set condition (Plate 13) is satisfied by an open set the shape of the inside of the fudgeflake (Figure 1.5.8); it may be thought of as the union of three terdragons.

Exercise 6.3.23. Made up of 6 parts, each with self-similarity for ratio list $(1/2, 1/2, 1/2)$.

Exercise 6.4.16. The hint is Plate 12.

Exercise 6.4.17. The boundary is made up of of sets A and B (Figure 6.4.11). Set B is made up of two parts similar to A, so dim A = dim B. Set A is made up of one part similar to A, and one part similar to B, which in turn is made up of two parts similar to A. So in all, A is made up of three parts similar to A. The ratio list is $(1/\sqrt{2}, 1/2\sqrt{2}, 1/2\sqrt{2})$.

Exercise 6.4.19. Figure 6.6.8.

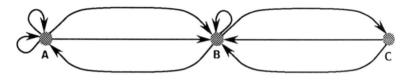

Figure 6.6.8. All edges have value $(3 - \sqrt{5})/2$.

Exercise 6.4.20. 1.22.

Exercise 6.4.21. [39].

Exercise 6.4.23. The graph of Exercise 4.3.12. This is a special case of the situation considered in [15].

Exercise 6.5.8. Consider the variant:

$$\widetilde{\mathcal{P}}^{s}_{\varepsilon}(F) = \sup \sum_{i} (2r_i)^s,$$

where the sup is over all countable disjoint collections $\{ B_{r_i}(x_i) : i \in \mathbb{N} \}$ with $x_i \in F$ and $2r_i < \varepsilon$. In $\mathbb{R}^d$, of course, diam $B_{r_i}(x_i) = 2r_i$, so this agrees with the definition in the text.

Exercise 6.5.17. Let S be a metric space and let $F \subseteq S$. The minimum number of balls $B_r(x)$ required to cover F will be called $M_r(F)$. The number $M_r(F) \cdot r^s$ is analogous to the number $\widetilde{\mathcal{K}}^{s}_{r}(F) = N_r(F) \cdot r^s$. If $S = \mathbb{R}^2$, then there are simple inequalities relating $M_r(F)$ and $N_r(F)$ showing that the critical index is the same. The definition using $M_r(F)$ makes sense in any metric space.

The fractile lines of the sandstone.
—*Scribner's Magazine*, April, 1893
(quoted in the *Oxford English Dictionary*)

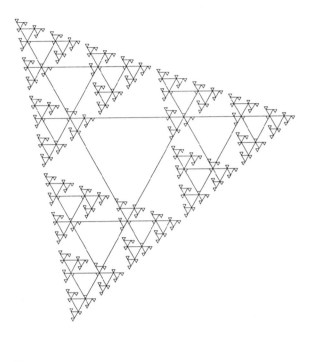

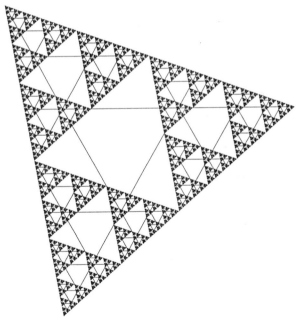

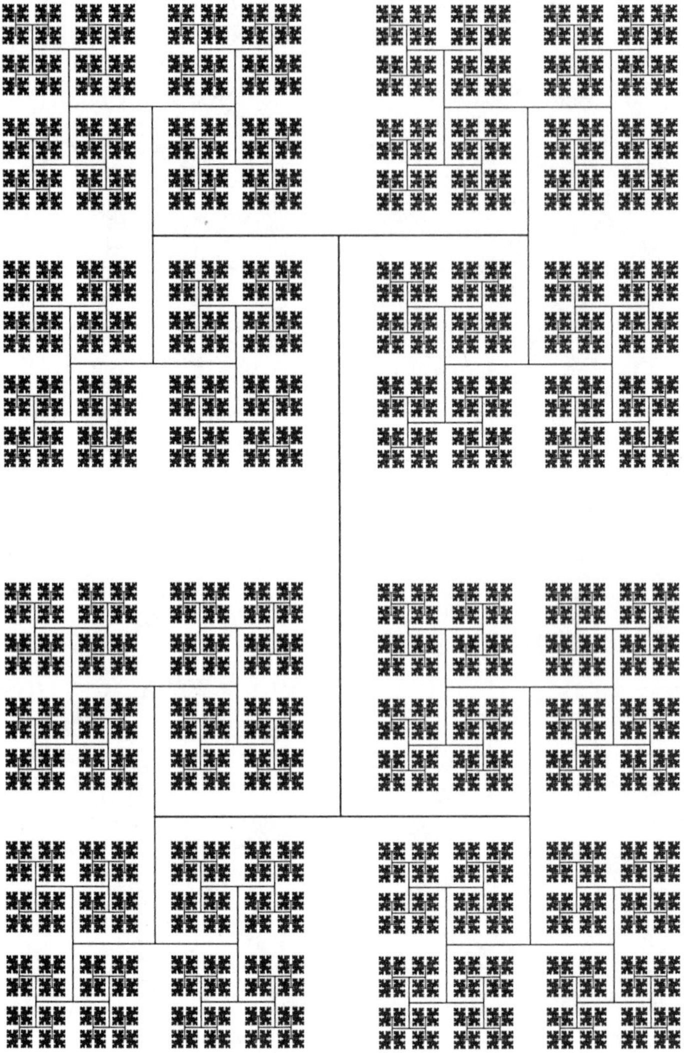

7

Additional Topics

This chapter includes additional examples of fractals, and hints at parts of the subject that we have not covered. In Sections 7.3 and 7.4, the computation of the fractal dimension requires more than just a simple application of the results of Chapter 6. So these examples show that there is more to the subject than we have seen in this book.

*7.1. A THREE-DIMENSIONAL DRAGON

Most of explicit examples that we have considered are sets in two-dimensional Euclidean space. One reason for this is that they are easier to visualize. But there is no mathematical reason not to go to higher dimensions. For example, three-dimensional dragon curves are possible.

Turtle graphics. A rotation in three-dimensional space is more complicated than in two-dimensional space. So turtle graphics in three dimensions requires more sophisticated kinds of turns.

Figure 7.1.1 shows a three-dimensional turtle from 6 sides. We can imagine the turtle keeping its local coordinate system (the **moving trihedron**) so that the origin is at the position of the turtle, the x-axis points forward, the y-axis points to the left, and the z-axis points upward. The three turning commands are: `roll`, `pitch`, and `yaw` (Figure 7.1.2). The turtle rotates about the x-axis for `roll`; the positive direction is "roll right". The turtle rotates about the y-axis for `pitch`; the positive direction is "pitch forward". The turtle rotates about the z-axis for `yaw`; the positive direction is "left".

Dragon. Here is a 3-dimensional dragon curve. Using right-angle turns, it produces helices within helices.

---

*An optional section.

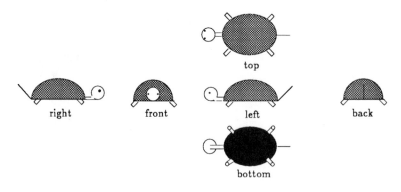

top

right   front   left   back

bottom

Figure 7.1.1. The turtle.

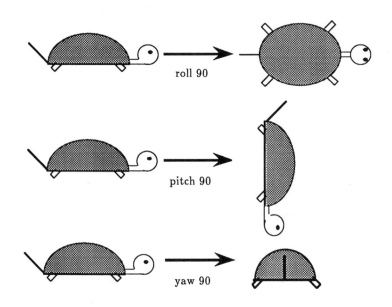

roll 90

pitch 90

yaw 90

Figure 7.1.2. Turns.

```
to drag :depth :size
    if :depth=0 [forward :size roll 90 stop]
    drag :depth-1 :size/3
    repeat 3 [yaw 90 drag :depth-1 :size/3]
    repeat 3 [yaw (-90) drag :depth-1 :size/3]
end
```

Figure 7.1.3. Dragon P_0.

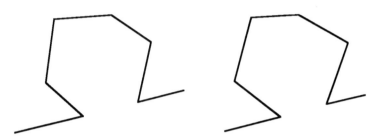

Figure 7.1.4. Dragon P_1.

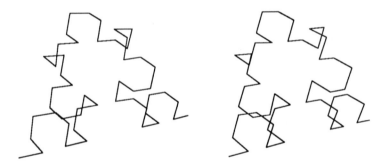

Figure 7.1.5. Dragon P_2.

In the illustrations (Figures 7.1.3 to 7.1.6), the turtle starts at the right and the faces toward the left foreground. They are shown in stereo pairs, drawn for the "cross-eye" method of viewing. The left eye looks at the right-hand picture, the right eye looks at the left-hand picture, and (somehow) the brain combines the two pictures into a three-dimensional image. It is important to have the picture horizontal; I find that bright illumination sometimes helps.

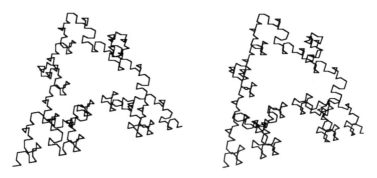

Figure 7.1.6. Dragon P_3.

(7.1.7) EXERCISE. *Prove that this sequence (P_k) converges, and the limit never crosses itself. (Therefore the limit set P has topological dimension 1.)*

(7.1.8) EXERCISE. *Compute the similarity dimension for the limit set P. Establish an open set condition. (Therefore the Hausdorff dimension of P is equal to the similarity dimension.)*

*7.2. OVERLAP

Next is a "similarity dimension" example with overlap. We will use the ratio list $(1/5, 1/5, 1/5)$. What is the dimension associated with it? Let a, b, c be three real numbers. Consider the three dilations f_1, f_2, f_3 of $\mathbb{R}$ with fixed points a, b, c, respectively. For certain choices of the points a, b, c, this realization satisfies the open set condition, and the invariant fractal K has Hausdorff dimension equal to the dimension of the ratio list. For certain other choices of a, b, c (such as two or three of them coincident) the Hausdorff dimension of K is not equal to the dimension of the ratio list. (There is, nevertheless, always an inequality, right?) What can we say in general?

Let us normalize things by assuming $a = 0$, $c = 1$, $0 < b < 1$. (Any choice of three distinct points can be reduced to this case.) What are the three functions f_j in this case? All three of the maps send $[0, 1]$ into itself, so the invariant set K is a subset of $[0, 1]$; in fact K may be constructed in the usual way by the contraction mapping theorem starting with $[0, 1]$.

(7.2.1) EXERCISE. *For what values of b are the three images of the open interval $(0, 1)$ disjoint?*

(7.2.2) EXERCISE. *Compute the Hausdorff dimension when $b = 1/10$.*

---

*An optional section.

(7.2.3) EXERCISE. *Compute the Hausdorff dimension when* $b = 1/5$.

There is a result of Falconer [20] that is relevant in situations like this. In this case it asserts that the Hausdorff dimension of the invariant set K is equal to the similarity dimension $\log 3/\log 5$ for *almost all* choices of $b \in [0, 1]$. That is, the set of all $b \in [0, 1]$ for which dim $K = \log 3/\log 5$ fails is a set of Lebesgue measure 0.

(7.2.4) EXERCISE. *Give an example of an iterated function system of similarities in* $\mathbf{R}^d$ *where the Hausdorff dimension of the invariant set coincides with the similarity dimension, but Moran's open set condition fails.*

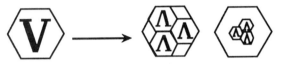

Figure 7.2.5. Similarities for Barnsley's wreath.

Barnsley's wreath. Barnsley's wreath is a fractal in the plane $\mathbf{R}^2$. There are six similarities of $\mathbf{R}^2$. Let V be a regular hexagon with side 1. The six maps, shown in Figure 7.2.5, constitute an iterated function system. The invariant set K is **Barnsley's wreath**, see Figure 7.2.6.

Figure 7.2.6. Barnsley's wreath.

The six images of the wreath overlap. So we get only an upper bound estimate of the dimension from this iterated function system: the solution of $3(1/2)^s + 3(1/4)^s = 1$ is

$$\frac{\log(3 + \sqrt{21})}{\log 2} - 1 \approx 1.9227;$$

so dim $K \leq$ Dim $K \leq 1.9227$. The exact dimension can be computed using graph self-similarity.

(7.2.7) EXERCISE. *Evaluate the Hausdorff dimension of Barnsley's wreath.*

*7.3. SELF-AFFINE SETS

The idea of an iterated function system makes good sense even when the maps are not similarities. One possibility that comes up often involves affine maps. The invariant set is then said to be **self-affine**. In the general self-affine case the evaluation of the Hausdorff dimension is not well understood. It has even been argued [37] that the Hausdorff dimension is not the proper dimension to use at all. We will present a few examples in this section.

A self-affine dust. As a reference, take the **unit square** in $\mathbf{R}^2$:

$$S = \{(x, y) : 0 \leq x \leq 1, 0 \leq y \leq 1\}.$$

The images of S under the two maps will be two rectangles:

$$R_1 = \{(x, y) : 0 \leq x \leq 1/2, 0 \leq y \leq 2/3\}$$
$$R_2 = \{(x, y) : 1/2 \leq x \leq 1, 1/3 \leq y \leq 1\}.$$

The function f_1 is an affine map of $\mathbf{R}^2$ onto itself, and sends the vertices of S to the corresponding vertices of R_1. The function f_2 is an affine map of $\mathbf{R}^2$ onto itself, and sends the vertices of S to the corresponding vertices of R_2.

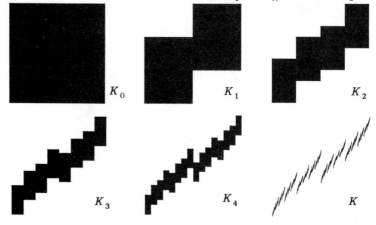

Figure 7.3.1. Self-affine dust.

---

*An optional section.

(7.3.2) EXERCISE. *There is a unique compact nonempty set $K \subseteq \mathbf{R}^2$ such that $K = f_1[K] \cup f_2[K]$.*

(7.3.3) EXERCISE. *Compute the Hausdorff dimension of the set K.*

Kiesswetter's curve. This is illustrated in two different ways. The set can be decomposed into four subsets, which are affine images of the whole thing. The four affine maps may be written in matrix notation. A point (x, y) in the plane is identified with a 2×1 column matrix.

$$f_1 \begin{bmatrix} x \\ y \end{bmatrix} = \begin{bmatrix} 1/4 & 0 \\ 0 & -1/2 \end{bmatrix} \begin{bmatrix} x \\ y \end{bmatrix}$$

$$f_2 \begin{bmatrix} x \\ y \end{bmatrix} = \begin{bmatrix} 1/4 & 0 \\ 0 & 1/2 \end{bmatrix} \begin{bmatrix} x \\ y \end{bmatrix} + \begin{bmatrix} 1/4 \\ -1/2 \end{bmatrix}$$

$$f_3 \begin{bmatrix} x \\ y \end{bmatrix} = \begin{bmatrix} 1/4 & 0 \\ 0 & 1/2 \end{bmatrix} \begin{bmatrix} x \\ y \end{bmatrix} + \begin{bmatrix} 1/2 \\ 0 \end{bmatrix}$$

$$f_4 \begin{bmatrix} x \\ y \end{bmatrix} = \begin{bmatrix} 1/4 & 0 \\ 0 & 1/2 \end{bmatrix} \begin{bmatrix} x \\ y \end{bmatrix} + \begin{bmatrix} 3/4 \\ 1/2 \end{bmatrix}.$$

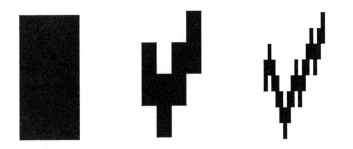

Figure 7.3.4. Kiesswetter's curve.

The first construction starts with the rectangle $L_0 = [0, 1] \times [-1, 1]$, and at each stage replaces the current set L_n with $L_{n+1} = f_1[L_n] \cup f_2[L_n] \cup f_3[L_n] \cup f_4[L_n]$. Because each of the maps f_j sends L_0 to a subset of L_0, this results in a decreasing sequence of compact sets. Kiesswetter's curve is the intersection $\bigcap_{n \in \mathbf{N}} L_n$.

The second construction starts with the line segment from $(0, 0)$ to $(1, 1)$, and makes the same transformation as before. Since $f_j\big((1, 1)\big) = f_{j+1}\big((0, 0)\big)$ for $j = 1, 2, 3$, these sets are all polygons. They are graphs of a sequence of continuous functions defined on $[0, 1]$; this sequence converges uniformly. The limit g is called **Kiesswetter's function.** Its graph $\{ (x, y) : y = g(x) \}$ is Kiesswetter's curve.

Figure 7.3.5. Kiesswetter's curve.

(7.3.6) EXERCISE. *Let g be Kiesswetter's function. Then for any integers $k \geq 0$ and $0 \leq j < 2^k$, prove*

$$\left| g\left(\frac{j}{4^k}\right) - g\left(\frac{j+1}{4^k}\right) \right| = \frac{1}{2^k}.$$

(7.3.7) EXERCISE. *Kiesswetter's curve is the graph of a continuous but nowhere differentiable function $g: [0,1] \to \mathbb{R}$.*

(7.3.8) EXERCISE. *Find the Hausdorff and packing dimensions of Kiesswetter's curve.*

Besicovitch-Ursell functions. Besicovitch and Ursell investigated the dimension of the graphs of non-differentiable functions. The most famous examples of these functions, dating back to Weierstrass, have a form

$$f(x) = \sum_{k=0}^{\infty} a_k \sin(b_k x),$$

for appropriate choices of a_k and b_k. A simpler variant was used by Besicovitch and Ursell, which will now be described.
Define a function $g: \mathbb{R} \to \mathbb{R}$ by:

$$g(x) = x \qquad \text{for } -1/2 \leq x \leq 1/2$$
$$g(x) = 1 - x \qquad \text{for } 1/2 \leq x \leq 3/2$$
$$g(x+2) = g(x) \qquad \text{for all } x.$$

If $0 < a < 1$, the **Besicovitch-Ursell** function with parameter a is:

$$f(x) = \sum_{k=0}^{\infty} a^k g(2^k x).$$

Partial sums $\sum_{k=0}^{n} a^k g(2^k x)$ of the series are illustrated in Figure 7.3.9, with $a = 0.6$. The pictures show only $0 \leq x \leq 1$, but the rest of the graph is simply related to this part.

Figure 7.3.9. Partial sums.

(7.3.10) EXERCISE. *The function $f(x)$ exists and is continuous.*

(7.3.11) EXERCISE. *Is the graph of f the invariant set for some iterated function system?*

Pictures for various values of a are shown in Figure 7.3.15 (page 204).

(7.3.12) EXERCISE. *For what values of a does the function f have bounded increase?*

(7.3.13) EXERCISE. *Compute the Hausdorff dimension of the graph of the Besicovitch-Ursell function with parameter $a = 3/5$.*

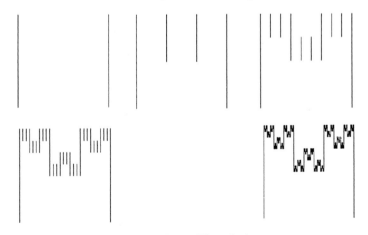

Figure 7.3.14. Hironaka's curve.

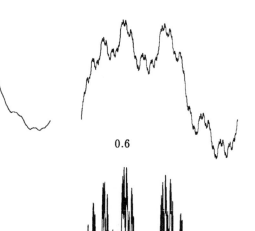

0.4   0.6

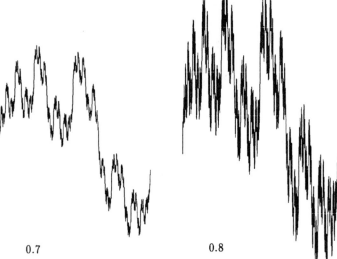

0.7   0.8

Figure 7.3.15. Besicovitch-Ursell functions.

Hironaka's curve. Pictured (Figure 7.3.14, page 203) are some approximations to **Hironaka's curve.** The first approximation consists of two vertical line segments, one unit long, one unit apart. For each subsequent approximation, additional line segments are added. The length of the new line segments is decreased by a factor of 1/2 at each stage. The distance between the line segments is decreased by a factor of 1/3 at each stage. The position of the line segments is determined by the pattern illustrated. Hironaka's curve is the limit set.

(7.3.16) EXERCISE. *Find topological and Hausdorff dimensions for Hironaka's curve.*

Number systems. Here is a way to generalize the "number systems" of Section 1.6. Elements of $\mathbf{R}^d$ should be identified with $d \times 1$ column matrices. Let D be a finite set in $\mathbf{R}^d$, including 0, and let B be a $d \times d$ matrix. What conditions should B satisfy so that all of the following vectors exist?

$$\sum_{j=1}^{\infty} B^j a_j,$$

where $a_j \in D$. The set F of all these vectors is the invariant set of an iterated function system of affine maps.

*7.4. OTHER EXAMPLES

There are some fractal sets in $\mathbf{R}$ that arise in connection with questions on the approximation of real numbers by rational numbers. If x is any irrational number, then it can be proved (for example, using continued fractions) that there are infinitely many rational numbers u/v with

$$\left| x - \frac{u}{v} \right| \le \frac{1}{v^2}.$$

But what happens if we want better approximation than that?

(7.4.1) EXERCISE. *Let $t > 2$ be an integer. Compute the Hausdorff dimension of the set*

$$K_t = \left\{ x : \left| x - \frac{u}{v} \right| \le \frac{1}{v^t} \text{ for infinitely many rational numbers } \frac{u}{v} \right\}.$$

An irrational number is a **surd** (or an algebraic number of degree 2) iff it is a solution of a quadratic equation with integer coefficients. Continued fractions can also be used to prove: If an irrational $x \in K_t$ for some $t > 2$, then x is not a surd.

Let us say that a real number x is a **Liouville number** iff for every integer $t > 0$, there exist infinitely many rational numbers u/v with

$$\left| x - \frac{u}{v} \right| \le \frac{1}{v^t}.$$

(7.4.2) EXERCISE. *Are there any irrational Liouville numbers?*

It can be proved that if x is an (irrational) Liouville number, then x is not an algebraic number at all; that is, x is not the solution of any polynomial equation with integer coefficients.

The **Appolonian gasket** is a subset of the plane $\mathbf{R}^2$ (see Figure 7.4.3). The first approximation is obtained by taking three mutually tangent circles with radius 1. The set C_0 is the region enclosed by three arcs (including the arcs themselves). Each approximation will consist of some regions bounded by three

---

*An optional section.

Figure 7.4.3. Appolonian gasket.

mutually tangent circular arcs. To obtain C_{k+1}, remove from each region of C_k the circle in the region tangent to all three of the arcs. (The boundary of the circle remains.) The Appolonian gasket is the "limit" (intersection) of the sets C_k.

The gasket is not self-similar, or even self-affine. But there are still features related to what we have seen before. If a function f (defined for complex numbers z) has the form

$$f(z) = \frac{az + b}{cz + d},$$

where $a, b, c, d \in \mathbb{C}$, then f is called a **linear fractional transformation**. An important property of a linear fractional transformation is that it maps circles to circles (provided a line is considered to be a circle of infinite radius).

(7.4.4) EXERCISE. *The Appolonian gasket is an invariant set for an iterated function system of linear fractional transformations.*

(7.4.5) EXERCISE. *Discuss the topological dimension and fractal dimension of the Appolonian gasket.*

A non-closed fractal. Most of the fractals that have been considered in this book are closed sets (or even compact sets). We will discuss now an example that is not closed. Mandelbrot calls it the **Besicovitch fractal**. It was studied by Besicovitch and Eggleston; more recently it occurs in the physics literature in connection with "multifractals" or "fractal measures". The proof will require some knowledge of probability theory, however.

Given $x \in [0, 1]$, consider its binary expansion, $x = \sum_{i=1}^{\infty} a_i 2^{-i}$, where each a_i is 0 or 1. We are interested in the frequency of the occurrence of the digit 0. More precisely, let $K_n^{(0)}(x)$ be the number of 0's and $K_n^{(1)}(x)$ the number of 1's occurring among the first n digits, $(a_1, a_2, \cdots, a_n)$. The **frequencies** in question are

$$F^{(0)}(x) = \lim_{n \to \infty} \frac{K_n^{(0)}(x)}{n},$$

$$F^{(1)}(x) = \lim_{n \to \infty} \frac{K_n^{(1)}(x)}{n}.$$

(Of course, the limits in question exist for only some $x \in [0, 1]$.)

Fix a number p, with $0 < p < 1$. We are interested in the set

$$S_p = \left\{ x \in [0,1] : F^{(0)}(x) \text{ exists, and } F^{(0)}(x) = p \right\}.$$

We will compute the Hausdorff dimension of the set S_p. If we write $q = 1 - p$, so that $F^{(0)}(x) = p$ implies $F^{(1)}(x) = q$, then we will show that

$$\dim S_p = \frac{-p \log p - q \log q}{\log 2}.$$

The proof will use a string model, as usual. But it will also use the "strong law of large numbers", an important result from probability theory.

Before we turn to the proof, let us consider the set S_p more carefully. Note that $[0,1]$ is not equal to $\bigcup_{0<p<1} S_p$, since the limit $F^{(0)}(x)$ does not exist for many x.

I will next prove that the set S_p is a Borel set. (This is the first example we have seen where measurability is not immediately obvious.) First, given a, b, n, the set

$$\left\{ x : a \le K_n^{(0)}(x) \le b \right\}$$

is a Borel set, since it consists of a finite number of intervals of length 2^{-n}. Then

$$S_p = \left\{ x : \lim_{n \to \infty} \frac{K_n^{(0)}(x)}{n} = p \right\}$$

$$= \bigcap_{k \in \mathbb{N}} \bigcup_{N \in \mathbb{N}} \bigcap_{n \ge N} \left\{ x : p - \frac{1}{k} \le \frac{K_n^{(0)}(x)}{n} \le p + \frac{1}{k} \right\},$$

so S_p is a Borel set.

Next, note that if x and y agree except in the first k digits, then $F^{(0)}(x) = F^{(0)}(y)$. So any open interval in $[0,1]$ intersects S_p. That is, S_p is dense in $[0,1]$. Certainly $S_p \neq [0,1]$, so of course S_p is not closed.

If the digits of x are all shifted to the right, and a new digit is added on the left, then the frequencies are unchanged. So S_p exhibits a natural self-similarity: If $x \in [0,1]$, then $F^{(0)}(x) = F^{(0)}(x/2) = F^{(0)}(1/2 + x/2)$. Thus the two similarities

$$f_0(x) = \frac{x}{2},$$

$$f_1(x) = \frac{x+1}{2},$$

have the property

$$S_p = f_0[S_p] \cup f_1[S_p],$$

with no overlap. The similarity dimension of the iterated function system (f_0, f_1) is 1. The conclusion is: similarity dimension may be misleading for non-closed sets.

(7.4.6) THEOREM. *The Hausdorff dimension of the set S_p is*

$$s = \frac{-p \log p - q \log q}{\log 2}.$$

PROOF: Let $E = \{0, 1\}$ be our two-letter alphabet, and recall the "base 2" model map $h \colon E^{(\omega)} \to [0, 1]$ defined on page 13. Then $h[E^{(\omega)}] = [0, 1]$. Also, $\operatorname{diam}[\alpha] = \operatorname{diam} h[[\alpha]] = 2^{-n}$ if $\alpha \in E^{(n)}$. We define frequencies for strings in the same way as for numbers: For $\alpha \in E^{(*)}$, let $K^{(0)}(\alpha)$ be the number of 0's in α, let $K^{(1)}(\alpha)$ be the number of 1's in α. Let

$$F^{(0)}(\sigma) = \lim_{n \to \infty} \frac{K^{(0)}(\sigma \restriction n)}{n},$$

$$F^{(1)}(\sigma) = \lim_{n \to \infty} \frac{K^{(1)}(\sigma \restriction n)}{n}.$$

These limits are defined for some strings $\sigma \in E^{(\omega)}$, and not for others. Let

$$T_p = \left\{ \sigma \in E^{(\omega)} : F^{(0)}(\sigma) = p \right\}.$$

Then clearly $S_p = h[T_p]$.

Now consider a measure $\mathcal{M}_p$ defined on $E^{(\omega)}$ as follows. Let $\alpha \in E^{(n)}$. If $k = K^{(1)}(\alpha)$, $n - k = K^{(0)}(\alpha)$, let $w_\alpha = p^{n-k} q^k$. Then $w_\alpha = w_{\alpha 0} + w_{\alpha 1}$, so these numbers define a metric measure $\mathcal{M}_p$ on $E^{(\omega)}$ with $\mathcal{M}_p([\alpha]) = w_\alpha$ for all $\alpha \in E^{(*)}$.

Now we require the result from probability theory. According to the measure $\mathcal{M}_p$ just defined, the "digits" of σ constitute independent Bernoulli trials, with probability p of outcome 0 and probability $q = 1 - p$ of outcome 1. So by the strong law of large numbers (for example, [6, Example 6.1]), we have

$$\mathcal{M}_p(T_p) = 1, \qquad \text{or, equivalently,} \qquad \mathcal{M}_p(E^{(\omega)} \setminus T_p) = 0.$$

We will take the case $p < 1/2$. The case $p > 1/2$ is similar, and the case $p = 1/2$ is the usual measure $\mathcal{M}_{1/2}$ and dimension 1 computed before (Proposition 6.2.1).

We begin with the upper bound, $\dim S_p \le s$. Let $\varepsilon > 0$ be given, and let $N \in \mathbb{N}$ satisfy $2^{-N} < \varepsilon$. Let $q' < q$. We will show that $\dim S_p \le s'$, where $s' = (-q' \log q - (1 - q') \log p)/\log 2$.

Consider the set $G \subseteq E^{(*)}$ defined as follows: if $\alpha \in E^{(n)}$, and $k = K^{(1)}(\alpha)$, then $\alpha \in G$ iff $k/n > q'$. For such α, we have $\operatorname{diam}[\alpha] = 2^{-n}$ and

$$\mathcal{M}_p([\alpha]) = p^{n-k} q^k = p^n \left(\frac{q}{p} \right)^k$$

$$> p^n \left(\frac{q}{p} \right)^{q' n} = p^{(1-q')n} q^{q' n}$$

$$= \left(2^{-n} \right)^{s'} = \left(\operatorname{diam}[\alpha] \right)^{s'}.$$

Let G' be the set of all $\alpha \in G$ with length $|\alpha| \geq N$ but $\alpha \restriction n \notin G$ for $N \leq n < |\alpha|$. That is, α belongs to G, but no ancestors of α (except possibly ancestors before generation N) belong to G. If $\sigma \in T_p$, then $\lim_{n \to \infty} K^{(1)}(\sigma \restriction n)/n = q > q'$, so for some $n \geq N$ we have $\sigma \restriction n \in G$, and therefore for some $n \geq N$ we have $\sigma \restriction n \in G'$. So

$$\{ [\alpha] : \alpha \in G' \}$$

is a disjoint cover of T_p. But

$$\sum_{\alpha \in G'} \left(\operatorname{diam} [\alpha] \right)^{s'} < \sum_{\alpha \in G'} \mathcal{M}_p ([\alpha])$$

$$= \mathcal{M}_p \left(\bigcup_{\alpha \in G'} [\alpha] \right) \leq 1.$$

Therefore $\overline{\mathcal{H}}_\varepsilon^{s'}(T_p) \leq 1$. Let $\varepsilon \to 0$ to conclude $\mathcal{H}^{s'}(T_p) \leq 1$, and therefore $\dim T_p \leq s'$. Now when $q' \to q$ we have $s' \to s$, so $\dim T_p \leq s$.

Now the model map h has bounded decrease, so $\dim S_p \leq s$. (Or, cover S_p with the sets $h[[\alpha]], \alpha \in G'$.)

Next I must prove the lower bound, $\dim S_p \geq s$. Let $q' > q$, and define $s' = (-q' \log q - (1 - q') \log p)/\log 2$. I will show $\dim S_p \geq s'$. Now

$$\mathcal{M}_p \left\{ \sigma \in E^{(\omega)} : \lim \frac{K^{(1)}(\sigma \restriction n)}{n} = q \right\} = \mathcal{M}_p(T_p) = 1,$$

and therefore

$$\mathcal{M}_p \left\{ \sigma : \text{there exists } N \in \mathbb{N} \text{ such that for all } n \geq N, \frac{K^{(1)}(\sigma \restriction n)}{n} < q' \right\}$$
$$= 1.$$

So by countable additivity,

$$\lim_{N \to \infty} \mathcal{M}_p \left\{ \sigma : \sup_{n \geq N} \frac{K^{(1)}(\sigma \restriction n)}{n} < q' \right\} = 1.$$

Choose N so that $\mathcal{M}_p(F) > 1/2$, where

$$F = \left\{ \sigma : \sup_{n \geq N} \frac{K^{(1)}(\sigma \restriction n)}{n} < q' \right\}.$$

Let $\varepsilon = 2^{-N}$.

Suppose $\mathcal{A}$ is a countable cover of S_p by sets A with $\operatorname{diam} A \leq \varepsilon$. First, we reduce to a cover by intervals of the form $h[[\alpha]]$. Each set $A \in \mathcal{A}$ is covered by (at most) three of the intervals $h[[\alpha]]$, where the length $|\alpha|$ is the integer n

with $2^{-n} < \text{diam } A \leq 2^{-n+1}$. Let $G \subseteq E^{(*)}$ be the set of all these α. (We may assume that the sets $[\alpha]$ are disjoint, since if two of them intersect, then one is a subset of the other, so we may delete the smaller one.) Thus

$$\sum_{\alpha \in G} \left(\text{diam}\,[\alpha]\right)^{s'} < 3 \sum_{A \in \mathcal{A}} (\text{diam } A)^{s'}$$

and $T_p \subseteq \bigcup_{\alpha \in G}[\alpha]$, so $\mathcal{M}_p\left(\bigcup_{\alpha \in G}[\alpha]\right) = 1$.

For $\alpha \in G$ we have $|\alpha| \geq N$. If $[\alpha] \cap F \neq \varnothing$, $|\alpha| = n$ and $K^{(1)}(\alpha) = k$, then

$$\mathcal{M}_p\left([\alpha]\right) = p^{n-k}q^k = p^n \left(\frac{q}{p}\right)^k$$

$$< p^n \left(\frac{q}{p}\right)^{q'n} = p^{(1-q')n}q^{q'n}$$

$$= \left(2^{-n}\right)^{s'} = \left(\text{diam}\,[\alpha]\right)^{s'}.$$

Now if $G' = \{\,\alpha \in G : [\alpha] \cap F \neq \varnothing\,\}$, then

$$\frac{1}{2} < \mathcal{M}_p(F) \leq \mathcal{M}_p\left(\bigcup_{\alpha \in G'} [\alpha]\right)$$

$$= \sum_{\alpha \in G'} \mathcal{M}_p\left([\alpha]\right) < \sum_{\alpha \in G'} \left(\text{diam}\,[\alpha]\right)^{s'}$$

$$\leq \sum_{\alpha \in G} \left(\text{diam}\,[\alpha]\right)^{s'} < 3 \sum_{A \in \mathcal{A}} (\text{diam } A)^{s'}.$$

Now $\mathcal{A}$ is any cover of S_p by sets of diameter $\leq \varepsilon$, so $\overline{\mathcal{H}}_\varepsilon^{s'}(S_p) > 1/6$. Therefore $\mathcal{H}^{s'}(S_p) > 1/6$, so dim $S_p \geq s'$. Now let $q' \to q$ to obtain dim $S_p \geq s$. ☺

(7.4.7) EXERCISE. *Compute the lower entropy index (the box dimension) of the set S_p.*

(7.4.8) EXERCISE. *Compute the packing dimension Dim S_p.*

(7.4.9) EXERCISE. *Let E be a finite alphabet, let $\mathcal{M}$ be a metric measure on the space $E^{(\omega)}$ of infinite strings, and let ρ be a metric on $E^{(\omega)}$. Suppose t is a positive real number, and let*

$$S = \left\{\, \sigma \in E^{(\omega)} : \lim_{n \to \infty} \frac{\log \mathcal{M}\left([\sigma \restriction n]\right)}{\log \text{diam}\,[\sigma \restriction n]} = t \,\right\}.$$

If $0 < \mathcal{M}(S) < \infty$, does it follow that dim $S = t$?

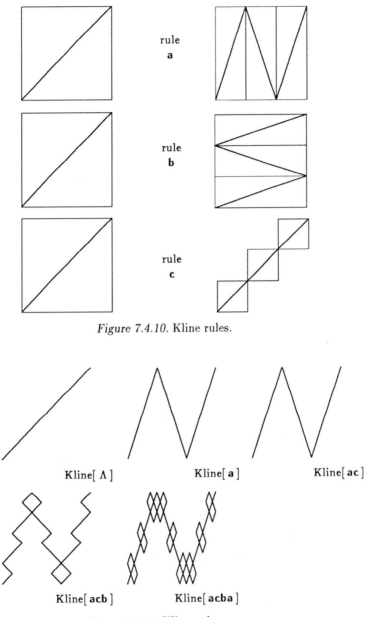

Figure 7.4.10. Kline rules.

Kline[Λ]   Kline[a]   Kline[ac]

Kline[acb]   Kline[acba]

Figure 7.4.11. Kline polygons.

Kline curves. The next example is a group of curves known as "Kline curves". They are subsets of the plane $\mathbb{R}^2$. We begin with the line segment from the point $(0,0)$ to the point $(1,1)$; it is the diagonal of the rectangle (actually a square) $[0,1] \times [0,1]$. There are three rules used to build the more complicated Kline polygons. Each of them replaces each of the line segments by three line segments. Rule **a** is implemented by subdividing the horizontal dimension of the containing rectangle in thirds, and replacing the diagonal by a three-part zig-zag, as illustrated. Rule **b** is implemented by subdividing the vertical dimension of the containing rectangle in thirds, and replacing the diagonal by a three-part zig-zag. Rule **c** is implemented by subdividing both the horizontal and vertical dimensions by three, and replacing the line segment by three parts of itself, inside the three diagonal subrectangles.

Each Kline polygon is obtained by applying these three rules in some order. Each finite string built from the alphabet $\{\mathbf{a}, \mathbf{b}, \mathbf{c}\}$ may be considered a "program" for the construction of a polygon. Several examples are illustrated. We will write Kline$[\alpha]$ for the Kline polygon corresponding to the string α.

Now let $\sigma \in \{\mathbf{a}, \mathbf{b}, \mathbf{c}\}^{(\omega)}$ be an infinite string. The **Kline curve** Kline$[\sigma]$ is the limit of the Kline polygons Kline$[\sigma \restriction k]$ as k increases.

(7.4.12) EXERCISE. *If σ is an infinite string from the alphabet $\{\mathbf{a}, \mathbf{b}, \mathbf{c}\}$, then* Kline$[\sigma \restriction k]$ *converges in the Hausdorff metric.*

What is the justification of the use of the word "curve"?

(7.4.13) EXERCISE. *Discuss the Hausdorff and packing dimensions of a Kline curve.*

*7.5. REMARKS

Michael Barnsley's wreath is from [**3**]. That text suggests that it is the invariant set for a contracting ratio list (r_1, r_2, r_3, r_4).

Karl Kiesswetter's curve is from [**33**]. It was proposed as a particularly elementary example of a continuous but nowhere differentiable function.

Theorem 7.4.6 is due to A. S. Besicovitch [**4**, Part II] and H. G. Eggleston [**17**]. Another proof is given by Patrick Billingsley [**5**, Section 14].

Dimension of a graph. The Besicovitch-Ursell functions (and the components of the Kline curves) satisfy Hölder conditions. A function $f \colon [a, b] \to \mathbb{R}$ satisfies a **Hölder condition** of order p iff there is a constant M such that

$$|f(x) - f(y)| \leq M|x - y|^p \qquad \text{for all } x, y \in [a, b].$$

(7.5.1) THEOREM. *Let $0 < p \leq 1$. Suppose $f \colon [a, b] \to \mathbb{R}$ satisfies a Hölder condition of order p. Then the graph*

$$G = \big\{\, (x, f(x)) : x \in [a, b] \,\big\}$$

satisfies dim $G \leq 2 - p$.

PROOF: We may assume that $[a, b] = [0, 1]$. Divide $[0, 1]$ into n sub-intervals of length $1/n$. On each of these intervals f can vary by no more than $M(1/n)^p$. Thus, the part of the graph over one of the sub-intervals can be covered by no more than $Mn^{1-p} + 1$ squares of side $1/n$. Thus

$$\overline{\mathcal{H}}^s_{\sqrt{2}/n}(G) \leq n(Mn^{1-p} + 1) \left(\frac{\sqrt{2}}{n}\right)^s = M2^{s/2}n^{2-p-s} + 2^{s/2}n^{1-s}.$$

If $s = 2 - p$, then this shows $\mathcal{H}^s(G) \leq M2^{(2-p)/2} + 2^{(2-p)/2}$, so dim $G \leq s = 2 - p$. ☺

Besicovitch and Ursell [4, part V] gave examples of functions satisfying a Hölder condition of order p (and no better) where dim $G = 2 - p$ and other examples where dim $G < 2 - p$.

Comments on the exercises. Exercise 7.2.2. The open set condition is satisfied, but not by the open set $(0, 1)$.

Exercise 7.2.3. The set K is contained in $[0, 1]$. Consider 4 parts of the set:

$$A = K \cap [4/5, 1]$$
$$B = K \cap [1/5, 4/5]$$
$$C = K \cap [4/25, 1/5]$$
$$D = K \cap [0, 4/25],$$

and observe that $(1/5)A \subseteq C$. Show that the graph similarity obeys Figure 7.5.2, and the open set condition is satisfied for the corresponding realization. The dimension is $-\log a/\log 5$, where $a \approx 0.3992$ is a solution of $a^4 - a^3 - a^2 + 3a - 1 = 0$. The dimension is approximately 0.57058; compare it to the upper bound obtained from the the ratio list $(1/5, 1/5, 1/5)$, namely $\log 3/\log 5 \approx 0.6826$.

Exercise 7.2.7. Hint: Use the method shown above for Exercise 7.2.3.

Exercise 7.3.7. If g is differentiable at a point a, and $x_n \leq a \leq y_n$, $\lim x_n = a = \lim y_n$, $x_n < y_n$, then

$$\lim_{n \to \infty} \frac{g(y_n) - g(x_n)}{y_n - x_n} = g'(a).$$

This is false by Exercise 7.3.6.

Exercise 7.3.8: $3/2$.

Exercise 7.3.13: [4, Part V].

Exercise 7.3.16: [40].

Exercise 7.4.1: [19, Section 8.5].

Exercise 7.4.5: [19, Section 8.4].

Exercise 7.4.9: [5, Section 14].

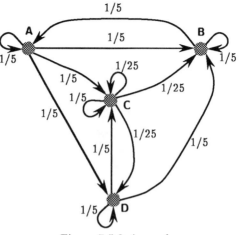

Figure 7.5.2. A graph.

Exercise 7.4.13: [**34**].

Exercise 7.2.7: The answer is dim $K = \text{Dim } K = \log t / \log 2 \approx 1.8459$, where $t \approx 3.5948$ is the largest zero of the polynomial

$$t^{13} - 5t^{12} + 3t^{11} + 5t^{10} + 5t^9 + 31t^8 - 62t^7$$
$$- 20t^6 + 14t^5 + 48t^4 - 60t^3 + 40t^2 - 48t + 48.$$

My graph has 13 nodes in it, and all edges have value $1/2$. (The number of nodes could be reduced by allowing edges with value $1/4$, or even $1/8$, but that would complicate the calculation of the spectral radius.) All 13 of the open sets for the open set condition are equilateral triangles. Can you find a simpler way to do this problem?

Life is a fractal in Hilbert space.
—Rudy Rucker, *Mind Tools*

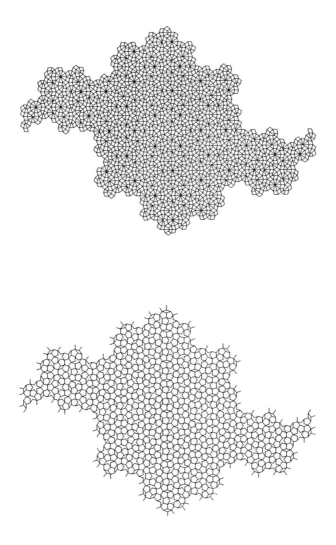

Appendix

Terms are listed here with the page number of an appropriate definition. In most cases, there is a reminder of the definition here, but for the complete definition see the page cited.

σ-algebra of sets, 132. Collection of subsets, contains the empty set and the whole space, closed under complements and countable unions.

accumulation point of a set, 42. Every ball centered at the point meets the set.

address, 14. Inverse image for the model map.

affine, 62. $f(tx + (1 - t)y) = tf(x) + (1 - t)f(y)$.

at most n-to-one, 98. The inverse image of any point consists of at most n points.

base for the open sets, 43.

bicompact, 55. Every family of closed sets with the finite intersection property has nonempty intersection.

Borel set, 133. Belongs to the σ-algebra generated by the open sets.

boundary point of a set, 49. Belongs to the closure of the set and to the closure of the complement.

boundary of a set, 49. The collection of all boundary points of the set.

bounded decrease, 51. $\rho(f(x), f(y)) \geq A \rho(x, y)$.

bounded distortion, 52. $A \rho(x, y) \leq \rho(f(x), f(y)) \leq B \rho(x, y)$.

bounded increase, 51. $\rho(f(x), f(y)) \leq B \rho(x, y)$.

Cauchy sequence, 48. $\rho(x_n, x_m) \to 0$.

clopen, 80. Closed and open.

closed ball, 41. $\overline{B_r}(x) = \{ y \in S : \rho(y, x) \leq r \}$.

closed set, 42. Contains all of its accumulation points.

closure of a set, 49. The set together with all of its accumulation points.

Notation

An index of the notations used in the book. See also page vii.

$\mathcal{L}^2$, two-dimensional Lebesgue measure: 139.

$\mathcal{L}^d$, d-dimensional Lebesgue measure: 140.

$\overline{\mathcal{H}}_\varepsilon^s$, Hausdorff outer measure: 147.

$\overline{\mathcal{H}}^s$, Hausdorff outer measure: 147.

$\mathcal{H}^s$, Hausdorff measure: 147.

dim S, Hausdorff dimension: 149.

$\mathcal{P}^s$, packing measure: 181.

Dim S, packing dimension: 182.

EXAMPLES

120-degree dragon: 22.
3-dimensional dragon: 196.
Appolonian gasket: 205.
Barnsley wreath: 199.
Besicovitch fractal: 206.
Besicovitch-Ursell function: 202.
Cantor dust: 1.
Eisenstein fractions: 32.
fudgeflake: 22.
golden rectangle fractal: 25.
Heighway dragon: 19.
Hironaka curve: 204.
I fractal: 167.
Kiesswetter curve: 201.

Kline curves: 206.
Koch curve: 18.
McWorter pentigree: 24.
Menger sponge: 109.
Peano curve: 65.
pentadendrite: 164.
Schmidt dragon: 167.
self-affine dust: 200.
Sierpiński gasket: 7.
snowflake curve: 19.
terdragon: 163.
twindragon: 30.
two-part dust: 167.

READING

The mathematical theory of fractal geometry, as I understand it, rests on three mathematical foundations: metric topology, measure theory, and probability. The metric topology in Chapter 2 is almost all that is generally needed for the study of fractal geometry. (Some additional "descriptive set theory" may be useful.) The measure theory of Chapter 5 is, however, only a part of what is needed for the deeper study of fractal geometry. Additional topics related to measure theory, such as integration, potential theory, and harmonic analysis, will be needed by the serious student of fractal geometry. (Some of the texts are [9], [29], [48].) Probability is an important branch of mathematics that is used often in fractal geometry. In order to reduce the background required here, I have almost entirely avoided using it explicitly. But further study of fractal geometry almost surely will require study of probability theory. And by "probability theory", I mean the modern subject that depends on measure theory, rather than the older version that gets by with calculus alone. (Some of the texts are [6], [8].)

Here are some suggestions for further reading on the topics considered in this book.

Benoit B. Mandelbrot, *The Fractal Geometry of Nature* [36]. This is the basic reference on fractals, together with a discussion of the applications of fractal sets in many branches of science. But it is not as mathematical as mathematicians would like, while it is too mathematical for many others. It contains many computer-generated pictures.

James Gleick, *Chaos: Making a New Science* [26]. This is a non-technical account by a New York *Times* reporter. It is concerned with the scientific phenomena governed by chaotic dynamical systems. "Chaos" and "fractals" are not the same field of study, but there are many relations between them.

Michael Barnsley, *Fractals Everywhere* [3]. This recent book is an introduction to fractal geometry from a somewhat different point of view than the one I have used here.

W. Hurewicz and H. Wallman, *Dimension Theory* [30]. This is a book on topological dimension. It is a bit out of date by now, but it has everything you need to know when you restrict attention to separable metric spaces. It requires background in metric topology, such as our Chapter 2. Other references for topological dimension are [18], [44] and [45].

K. J. Falconer, *The Geometry of Fractal Sets* [19]. This is a modern text on Hausdorff dimension. It contains much more material than I have included here.

C. A. Rogers, *Hausdorff Measures* [47]. This book is concerned with the more technical aspects of the subject. It was written before Mandelbrot's sudden popularity, illustrating the fact that mathematicians had been doing this sort of thing ever since Hausdorff. But before Mandelbrot, scientists (including many mathematicians) considered it to be of only minor importance.

H.-O. Peitgen and P. H. Richter, *The Beauty of Fractals* [46]. This book is a discussion of Julia sets, the Mandelbrot set, and related topics, mostly without proofs. It contains many computer-generated color pictures; also an essay on whether they should be considered to be "art".

Chandler Davis and Donald J. Knuth, "Number representations and dragon curves" [11]. Two papers discuss the Heighway dragon (and some other dragons), and how they are related to representations of numbers in complex bases.

Michel Dekking, Michel Mendès France, and Alf van der Poorten, "Folds!" [12]. (The title was inspired by a movie current at the time.) Heighway's dragon can be generated by folding a strip of paper. A reader who can get past the "humor" will learn about this, and many other interesting topics.

William A. McWorter and Jane M. Tazelaar, "Creating Fractals" [41]. Instructions for drawing a wide variety of dragon curves, with code in BASIC. This is the source for McWorter's pentigree.

REFERENCES

1. Harold Abelson and Andrea diSessa, *Turtle Geometry: The Computer as a Medium for Exploring Mathematics* (MIT Press, 1981).

2. M. F. Barnsley and S. Demko, "Iterated function systems and the global construction of fractals". *Proc. R. Soc. London* A **399** (1985) 243–275.

3. M. F. Barnsley, *Fractals Everywhere* (Academic Press, 1988).

4. A. S. Besicovitch, "Sets of fractional dimensions", Part I: *Math. Ann.* **101** (1929) 161–193; Part II: *Math. Ann.* **110** (1934) 321–329; Part III: *Math. Ann.* **110** (1934) 331–335; Part IV: *J. London Math. Soc.* **9** (1934) 126–131; Part V: *J. London Math. Soc.* **12** (1934) 18–25.

5. Patrick Billingsley, *Ergodic Theory and Information* (John Wiley & Sons, 1965).

6. Patrick Billingsley, *Probability and Measure* (John Wiley & Sons, 1979).

7. L. M. Blumenthal and K. Menger, *Studies in Geometry* (Freeman, 1970).

8. Kai Lai Chung, *A Course in Probability* (Academic Press, 1974).

9. Donald L. Cohn, *Measure Theory* (Birkhäuser, 1980).

10. H. S. M. Coxeter, *Introduction to Geometry* (John Wiley & Sons, 1961).

11. Chandler Davis and Donald J. Knuth, "Number representations and dragon curves". Part I: *J. Recreational Math.* **3** (1970) 66–81; Part II: *J. Recreational Math.* **3** (1970) 133–149.

12. Michel Dekking, Michel Mendès France, and Alf van der Poorten, "Folds!" Part I: *Math. Intelligencer* **4** (1982) 130–138; Part II: *Math. Intelligencer* **4** (1982) 173–181; Part III: *Math. Intelligencer* **4** (1982) 190–195.

13. F. M. Dekking, "Recurrent sets". *Advances in Math.* **44** (1982) 78–104.

14. Robert L. Devaney, *An Introduction to Chaotic Dynamical Systems* (Benjamin/Cummings Publishing Company, 1986).

15. Vladimir Drobot and John Turner, "Hausdorff dimension and Perron-Frobenius Theory". *Ill. J. Math.* **33** (1989) 1–9.

16. H. G. Eggleston, *Convexity* (Cambridge University Press, 1958).

17. H. G. Eggleston, "The fractional dimension of a set defined by decimal properties." *Quarterly J. Math.* **20** (1949) 31–36.

18. R. Engelking, *Dimension Theory* (North-Holland, 1978).

19. K. J. Falconer, *The Geometry of Fractal Sets* (Cambridge University Press, 1985).

20. K. J. Falconer, "The Hausdorff dimension of some fractals and attractors of overlapping construction". *J. Statist. Phys.* **47** (1987) 123–132.

21. K. J. Falconer, "The Hausdorff dimension of self-affine fractals". *Math. Proc. Camb. Phil. Soc.* **103** (1988) 339–350.

22. F. R. Gantmacher, *Matrix Theory* (Chelsey, 1959). Volume II, Chapter XIII: "Matrices with non-negative elements".

23. Martin Gardner, "Mathematical Games". *Scientific American* (March, April, and July, 1967).

24. William J. Gilbert, "Fractal geometry derived from complex bases". *Math. Intelligencer* **4** (1982) 78–86.

25. William J. Gilbert, "The fractal dimension of sets derived from complex bases". *Canad. Math. Bull.* **29** (1986) 495–500.

26. James Gleick, *Chaos: Making a New Science* (Viking, 1987).

27. F. Hausdorff, "Dimension und äußeres Maß". *Math. Ann.* **79** (1918) 157–179.

28. F. Hausdorff, *Mengenlehre*, 3 Auflage. (Dover Publications, 1944).

29. Edwin Hewitt and Karl Stromberg, *Real and Abstract Analysis* (Springer-Verlag, 1965).

30. W. Hurewicz and H. Wallman, *Dimension Theory* (Princeton University Press, 1941).

31. John E. Hutchinson, "Fractals and self similarity". *Indiana Univ. Math. J.* **30** (1981) 713–747.

32. J. L. Kelley, *General Topology* (Van Nostrand, 1955).

33. Karl Kießwetter, "Ein einfaches Beispiel für eine Funktion, welche überall stetig und nicht differenzierbar ist". *Math. Phys. Semesterber.* **13** (1966) 216–221.

34. S. A. Kline, "On curves of fractional dimensions". *J. London Math. Soc.* **20** (1945) 79–86.

35. K. Kuratowski, *Topology*, Volume I (Academic Press, 1966).

36. Benoit B. Mandelbrot, *The Fractal Geometry of Nature* (W. H. Freeman and Company, 1982).

37. Benoit B. Mandelbrot, "Self-affine fractal sets". In: *Fractals in Physics*, L. Pietronero and E. Tosatti, editors (Elsevier Science Publishers, 1986).

38. Jacques Marion, "Mesure de Hausdorff d'un fractal à similitude interne". *Ann. Sc. Math. Québec* **10** (1986) 51–84.

39. R. Daniel Mauldin and S. C. Williams, "Hausdorff dimension in graph directed constructions". *Trans. Amer. Math. Soc.* **309** (1988) 811–829.

40. Kurt McMullen, "The Hausdorff dimension of general Sierpiński carpets". *Nagoya Math. J.* **96** (1984) 1–9.

41. William A. McWorter and Jane M. Tazelaar, "Creating Fractals". *Byte* (August 1987) 123–134.

42. P. A. P. Moran, "Additive functions of intervals and Hausdorff measure". *Proc. Cambridge Phil. Soc.* **42** (1946) 15–23.

43. Frank Morgan, *Geometric Measure Theory* (Academic Press, 1988).

44. K. Nagami, *Dimension Theory* (Academic Press, 1970).

45. A. Pears, *Dimension Theory of General Spaces* (Cambridge University Press, 1975).

46. H.-O. Peitgen and P. H. Richter, *The Beauty of Fractals* (Springer-Verlag, 1986).

47. C. A. Rogers, *Hausdorff Measures* (Cambridge University Press, 1970).

48. R. L. Royden, *Real Analysis* (Macmillan, 1968).

49. S. James Taylor and Claude Tricot, "Packing measure, and its evaluation for a Brownian path". *Trans. Amer. Math. Soc.* **288** (1985) 679–699.

50. S. James Taylor, "The measure theory of random fractals". *Math. Proc. Camb. Phil. Soc.* **100** (1986) 383–406.

INDEX

See also the index of terms (page 217), the index of notation (page 220), and the list of examples (page 221).

Undergraduate Texts in Mathematics